Springer Briefs in Interdisciplinarv Geosciences – Asia-Pacific

Series Editor

Professor Yoshiharu Omura, University of Kyoto, Japan, Uji, Japan

The book series is a collection of monographs summarizing the advances in geosciences and emerging interdisciplinary research works as presented at the Asia Oceania Geosciences Society (AOGS) Annual Meetings. While papers in scientific journals such as Geoscience Letters are reports of new scientific insights and original modeling and analysis methods for specific phenomena and geophysical processes, it is very useful to publish summaries of research progress and state-of-the-art reviews of relevant publications to provide a comprehensive understanding of important problems and critical issues in geosciences. The book series will be available in both online e-book and hard-copy format for a fast dissemination of knowledge in geosciences to a wide range of readers.

Umesh Chandra Kulshrestha ·
Md. Mizanur Rahman · Arti Bhatia

Atmosphere-Biosphere Interactions of Reactive Nitrogen in South Asia

Air Pollution and Food Productivity Concerns

Umesh Chandra Kulshrestha
School of Environmental Sciences
Jawaharlal Nehru University
New Delhi, Delhi, India

Md. Mizanur Rahman
Gazipur Agricultural University
Gazipur, Bangladesh

Arti Bhatia
Division of Environmental Science
ICAR-Indian Agricultural Research Institute
New Delhi, Delhi, India

ISSN 2731-8923 ISSN 2731-8931 (electronic)
Springer Briefs in Interdisciplinary Geosciences – Asia-Pacific
ISBN 978-3-032-17193-1 ISBN 978-3-032-17194-8 (eBook)
https://doi.org/10.1007/978-3-032-17194-8

This Springer imprint is published by the registered company Springer Nature Switzerland AG
The registered company address is: Gewerbestrasse 11, 6330 Cham, Switzerland

Series Editor's Preface

Since 2004, the Asia Oceania Geosciences Society (AOGS) has held annual meetings where geoscientists, primarily from the Asia–Oceania region, gather to discuss a wide range of topics across eight sections: Atmospheric Sciences, Biogeosciences, Hydrological Sciences, Interdisciplinary Geosciences, Ocean Sciences, Planetary Sciences, Solar and Terrestrial Sciences, and Solid Earth Sciences. Leveraging the breadth and diversity of geosciences represented within the AOGS community, the Interdisciplinary Geosciences section has actively promoted collaboration among scientists from different fields, with a particular focus on the sustainable management of environmental challenges such as air pollution, water contamination, and climate change. Among the three founding members of AOGS—Ian Axford, Yohsuke Kamide, and Wing-Huen Ip. Prof. Ip has played a pivotal role in advancing interdisciplinary geosciences. As Chair of the Regional Advisory Committee, he has championed regional satellite workshops and seminars on environmental issues, furthering the AOGS vision: "*In Asia for Asia and the World*." As a first outcome of these efforts, we are pleased to announce the publication of the inaugural volume in the book series *SpringerBriefs in Interdisciplinary Geosciences—Asia Pacific*, titled: *Atmosphere-Biosphere Interactions of Reactive Nitrogen in South Asia: Air Pollution and Food Productivity Concerns.*

This book highlights the critical importance of managing nitrogen (N), one of the most essential nutrients for plant growth and a cornerstone of global food production. At the same time, nitrogen can harm the environment when released as reactive nitrogen species (RNS). As described in the chapters, RNS contribute to air pollution, soil acidification, eutrophication, and global warming through multiple pathways.

Each chapter has been carefully reviewed by distinguished experts:

- Kazuhide Matsuda
- John Philip Matthews
- James Galloway
- Van-Thanh-Van Nguyen
- Wing-Huen Ip

Their valuable contributions in enhancing the quality of the manuscripts are deeply appreciated.

Uji, Japan

Yoshiharu Omura
Editor-in-Chief
SpringerBriefs in Interdisciplinary
Geosciences—Asia Pacific

Acknowledgements We gratefully acknowledge the invaluable guidance of Prof. Mark Sutton, CEH Edinburgh, and Director of the UKRI GCRF SANH project. We also extend our sincere thanks for the financial support provided through the UKRI GCRF SANH project, which enabled the execution of new measurements. The moral support offered by the administrations of JNU, GAU, and ICAR-IARI during the preparation of this book is deeply appreciated. We are also thankful for the contributions of our students and colleagues, whose data and insights greatly enriched this work.

We sincerely thank Prof. Ip Wing-Huen, Chair, Regional Advisory Committee, Asia Oceania Geosciences Society for his constant encouragements to develop a book for the AOGS Interdisciplinary Geosciences-Asia-Pacific series. We also thank Professor Yoshiharu Omura, Series Editor-in-chief for including this book as part of Springer Briefs in Interdisciplinary Geosciences-Asia-Pacific.

Umesh Chandra Kulshrestha
Md. Mizanur Rahman
Arti Bhatia

Competing Interests The authors have no competing interests to declare that are relevant to the content of this manuscript.

Contents

Abbreviations

AAI	Ammonium Availability Index
AMO	Ammonia Monooxygenase
AN	Ammonium Nitrate
AOA	Ammonia-Oxidizing Archaea
AOB	Ammonia-Oxidizing Bacteria
ATS	Ammonium Thiosulfate
AWD	Alternate Wetting and Drying
BNF	Biological Nitrogen Fixation
BRDN	Biochar 2 t ha^{-1} + RDN
CBNF	Cultivated Biological Nitrogen Fixation
CCC	Coated Calcium Carbide
CDSupn	Cowdung 2 t ha^{-1} + supplemented N
CEC	Cation Exchange Capacity
CP	Chlorinated Pyridine
CPCB	Central Pollution Prevention and Control Board
CRFs	Controlled-Release Fertilizers
CRI	Crown Root Initiation
DAP	Diammonium Phosphate
DCD	Dicyandiamide
DIN	Dissolved Inorganic Nitrogen
DIs	Dual Inhibitors
DMPP	3,4-dimethylpyrazole-phosphate
DON	Dissolved Organic Nitrogen
DSR	Direct Seeded Rice
EEFs	Enhanced Efficiency Fertilizers
EU	European Union
FAO	Food and Agriculture Organization of the United Nations
FYM	Farmyard Manure
GAU	Gazipur Agricultural University
GHG	Greenhouse Gas
GWP	Global Warming Potential

HBNF	Haber-Bosch Nitrogen Fixation
IBDU	Isobutylidenediurea
IPCC	Intergovernmental Panel on Climate Change
ISRO	Indian Space Research Organization
LCCN	Leaf Color Chart-Based N
LULC	Land Use and Land Cover
MICPU	Maleic-Itaconic Copolymer-Treated Urea
NAAQS	National Ambient Air Quality Standards
NBPT	N-(n-butyl) thiophosphorictriamide
NCR	National Capital Region
Nis	Nitrification inhibitors
NPK	Nitrogen, Phosphorus and Potassium
NPPT	N-(n-propyl) thiophosphorictriamide
NUE	Nitrogen Use Efficiency
PCU/PGU	Polymer-Coated Urea/Phosphogysum Coated Urea
RDN	Recommended Dose of N
RNS	Reactive Nitrogen Species
SA	South Asia
SACEP	South Asia Cooperative Environment Programme
SANH	South Asian Nitrogen Hub
SCU	Sulfur-Coated Urea
SDGs	Sustainable Development Goals
SPADN	SPAD Meter-Based N
SRDI	Bangladesh Soil Resource Development Institute (SRDI)
SRDIAppN	App-Based N
SRF	Slow-Release Fertilizer
SRI	System of Rice Intensification
SSM	Sustainable Soil Management
STBN	Soil Test-Based N
TIN	Total Inorganic Nitrogen
TN	Total Nitrogen
UAN	Urea-Ammonium Nitrate
UK	United Kingdom
UN	United Nations
USG	Urea Super Granules
VOCs	Volatile Organic Compounds
WFPS	Water-Filled Pore Space

List of Figures

List of Tables

Chapter 1
Nitrogen Pollution

Nitrogen (N_2) gas makes up approximately 78% of the Earth's atmosphere, making it the most abundant gas in the atmosphere. Despite this fact, molecular nitrogen is largely biologically inaccessible to most organisms due to its inert nature. However, certain natural and anthropogenic processes convert the inert N_2 into its reactive forms as shown in Fig. 1.1. Major steps of N cycle include: (i) Nitrogen fixation, (ii) Nitrification, (iii) Assimilation, (iv) Ammonification and, (v) Denitrification. Nitrogen Fixation **c**onverts inert N_2 into reactive nitrogen species (RNS).Natural Fixation is performed by certain bacteria (e.g., *Rhizobium, Azotobacter*) in the soil or in symbiosis with leguminous plants where N_2 is converted to ammonia (NH_3). Anthropogenic fixationis achieved industrially via the Haber–Bosch process to create fertilizers. Abioticfixation N_2 is done when lightning strikes break the N_2 molecules and form nitrate (NO_3^-). In nitrification process, ammonia (NH_3) is converted to nitrite (NO_2^-) by *Nitrosomonas* bacteria. Then nitrite (NO_2^-) is oxidized to nitrate (NO_3^-) by *Nitrobacter*. During assimilation process, plants absorb nitrate (NO_3^-) or ammonium (NH_4^+) through their roots which are then incorporated into amino acids, proteins, and nucleic acids in plants. These are then transferred to animals through consumption. Ammonification includes decomposition of organic matter (from dead plants and animals, or waste).Decomposing microbes convert organic nitrogen compounds back into ammonium (NH_4^+). Denitrification is performed by anaerobic bacteria (e.g., *Pseudomonas, Clostridium*). In this process, nitrate (NO_2^-) is reduced back to nitrogen gas (N_2) or nitrous oxide (N_2O), which is released into the atmosphere. These transformations of N cycle allow nitrogen to circulate between the atmosphere, biosphere, and lithosphere. Without this cycle, the nitrogen locked in the atmosphere would remain biologically unavailable, despite being all around us.

Both natural and anthropogenic sources contribute RNS. Among natural sources of N, biological nitrogen fixation (BNF) and lightning are the major contributors, while among anthropogenic sources, fossil fuel emissions, cultivated biological N fixation and Haber–Bosch fixation are the major contributors. N Budgets from natural

U. C. Kulshrestha et al., *Atmosphere-Biosphere Interactions of Reactive Nitrogen in South Asia*, Springer Briefs in Interdisciplinary Geosciences – Asia-Pacific,
https://doi.org/10.1007/978-3-032-17194-8_1

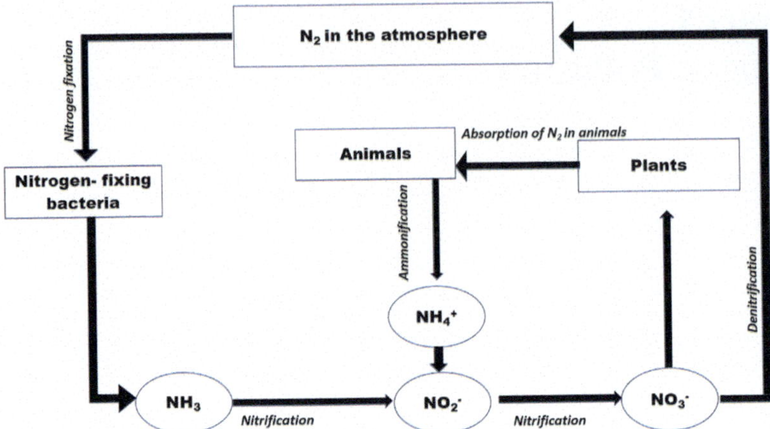

Fig. 1.1 Nitrogen cycle

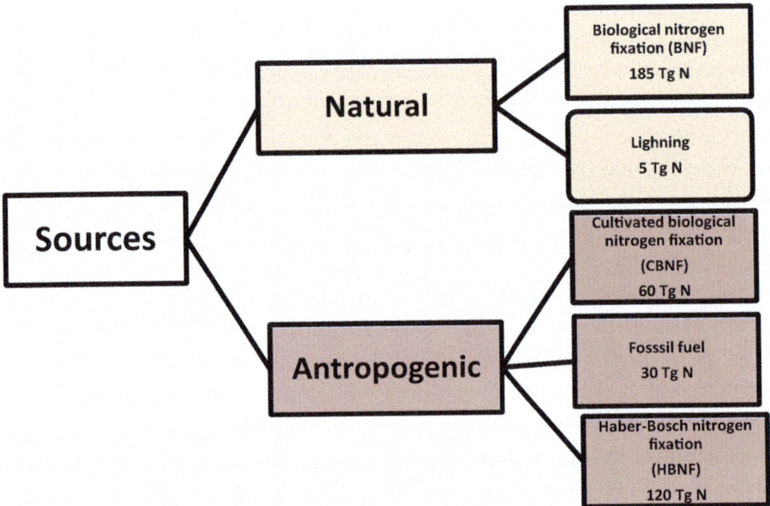

Fig. 1.2 N budgets from natural and anthropogenic sources

and anthropogenic sources and budgets based on Galloway et al. (2008) are shown in Fig. 1.2.

Nitrogen dioxide (NO_2), nitric oxide (NO), nitric acid (HNO_3), nitrous acid (HONO), nitrate (NO_3^-), nitrous oxide (N_2O), peroxyacetyl nitrate (PAN), ammonia (NH_3), ammonium (NH_4^+) and organic nitrogen(R-NH_2) are some examples of RNS (Fig. 1.3).

Figure 1.4 shows environmental impacts of RNS. The RNS are negatively impacting terrestrial and aquatic ecosystems polluting the air, water and soils, with

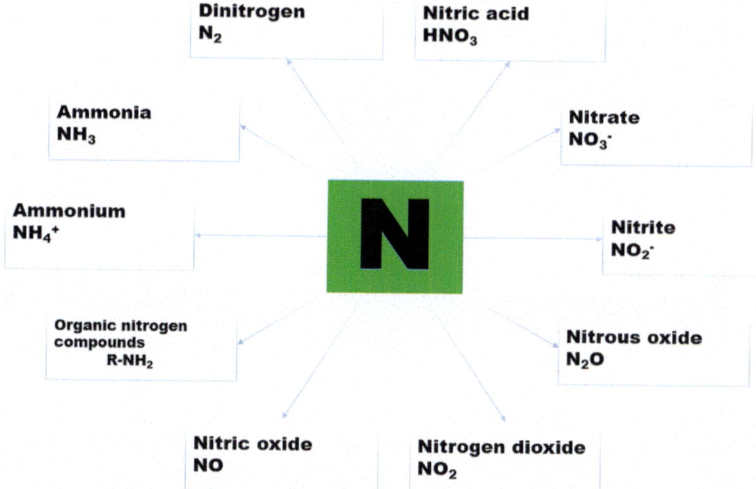

Fig. 1.3 Different reactive nitrogen species (RNS)

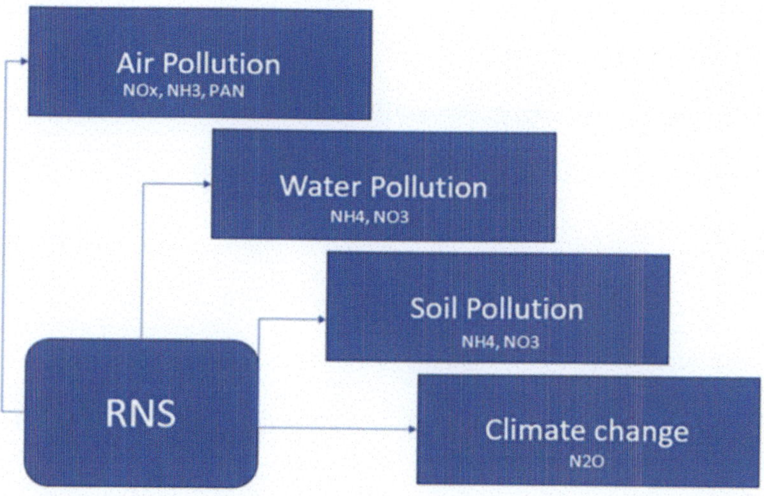

Fig. 1.4 RNS related environmental issues

several of them being major contributors to climate change, showing their importance for greenhouse balance. Nitrous oxide (N_2O), is a potent greenhouse gas which has approximately 300 times the global warming potential of CO_2 over a 100-year period.

Excess deposition of nitrogen compounds in soils can lead to nutrient imbalances, reducing plant biodiversity. RNS can acidify soils, impairing microbial communities and altering nutrient cycling. Elevated nitrogen levels from agricultural runoff and industrial waste cause eutrophication. An over-enrichment of water bodies that leads to algal blooms. notably Emissions arise from fertilized soils, wastewater, and combustion processes. Nitrogen oxides (NO_x) indirectly contribute to warming by participating in tropospheric ozone formation which is also a greenhouse gas. Nitrates in drinking water are linked to blue baby syndrome and potential long-term effects like cancer. Urban populations are especially vulnerable due to air emissions from transport and industry.

1.1 Natural Sources of RNS

Lightning strikes, volcanic activity, soil microbial activity, oceans and biological decay are some natural processes which release reactive nitrogen species (RNS). Anthropogenic activities mostly combustion processes which involve air as an oxidative medium convert N_2 into RNS.

Primary Reactions of Lightning-Generated NO_x are given below

$$NO + O_3 \rightarrow NO_2 + O_2 \tag{1.1}$$

NO reacts with ozone, forming NO_2 and releasing molecular oxygen.

$$NO_2 + \text{sunlight(hv)} \rightarrow NO + O \tag{1.2}$$

Sunlight photolyzes NO_2 back into NO and atomic oxygen.

$$O + O_2 \rightarrow O_3 \tag{1.3}$$

Atomic oxygen quickly reacts with molecular oxygen to regenerate ozone.

$$NO_x + VOCs + \text{sunlight} \rightarrow O_3 + \text{secondary pollutants} \tag{1.4}$$

In the presence of volatile organic compounds (VOCs), NO_x drives the formation of ground-level ozone and other oxidants.

1.2 Anthropogenic Sources of RNS

Fossil fuel combustion is a major anthropogenic process of generating RNS. Around 80% of the world's primary energy supply still comes from fossil fuels namely coal, oil, and natural gas. From 1965 to 2021, global coal consumption surged from 16,177 to 44,600 TWh, oil from 18,012 to 51,530 TWh, and natural gas from 6,303 TWh to 40,239 TW coal consumption. In South Asia, traditional biomass sources like dung cakes, crop residues, fuel wood, and other organic materials also remain

Table 1.1 Coal consumption of different countries

Country	Coal consumption (million tons)
China	4319.921
India	966.288
United States	731.071
Pakistan	10.199
Sri Lanka	2.295
Bangladesh	2.099
Afghanistan	1.871
Nepal	0.2283
Bhutan	0.096

Source (https://www.worldometers.info/coal/coal-consumption-by-country/)

widely used, particularly in rural communities. In India, about 41% of the population continues to rely on biomass for cooking. This reliance contributes approximately 340 million tonnes of CO_2 annually—roughly 13% of the country's total greenhouse gas emissions. Coal consumption of different countries is given in Table 1.1.

1.3 Nitrogen Oxides (NOx)

Nitrogen oxides are the most important RNS. Automobiles including cars, trucks, buses, and motorcycles release NO and NO_2 commonly known as NOx during fuel combustion. Nitrogen dioxide (NO_2) is a major reactive nitrogen species and a key pollutant under the National Ambient Air Quality Standards (NAAQS) in India. 40 μgm^{-3} limit is prescribed for annual NO_2 exposure and 80 μgm^{-3} for 24 h NO_2 exposure as safe limits for urban and industrial locations in India (CPCB 2009). Global trends from 2005 to 2018 show an increase in NO_2 levels, though regional variations exist. For instance, Eastern USA, Western Europe, Japan, and parts of China exhibit declining trends, while India, parts of China, and the Middle East display significant increases in NO_2 concentrations. In South Asia, NOx emissions rose dramatically between 1990 and 2017- Afghanistan increased from 16.19 to 44.96, Bangladesh from 120.24 to 531.12, Bhutan from 1.88 to 6.1, Maldives from 0.67 to 12.59, Nepal from 27.75 to 100.19, Pakistan from 433.13 to 1006.79, Sri Lanka from 45.35 to 179.81, and India from 3439.34 to 9673.33 Gg (Hooper et al. 2025).

Table 1.2 gives number of cars per capita for major cities in South Asian countries. However, per capita vehicle ownership might seem like a simple proxy for pollution, but it doesn't capture the full story. It depends on factors such as type of vehicle, fuel, usage patterns etc. A city with more heavy-duty diesel trucks emits far more pollutants than one with the same number of electric cars. Two cities may have the

Table 1.2 Per capita number of cars in different major cities of South Asia

City	Country	10^{-3} cars per capita
Colombo	Sri Lanka	1101.21
New Delhi	India	65.99
Kathmandu	Nepal	54.44
Karanchi	Pakistan	16.33
Dhaka	Bangladesh	0.90

Table 1.3 NO_2 levels at urban and rural sites in India

Site	$NO_2 \mu gm^{-3}$	References
Okhla, Delhi	24.4	Singh and Kulshrestha (2014)
Mai village	18.8	Singh and Kulshrestha (2014)

same per capita vehicle count, but vastly different emission profiles. A vehicle used for long daily commutes will contribute more to pollution than one used occasionally. Public transport reliance versus private use skews real-world emissions despite similar ownership rates.

Older or poorly maintained vehicles emit more NO_x, CO, $PM_{2.5}$ and VOCs. Stricter emission norms (Bharat Stage VI in India) reduce pollution per vehicle, even if numbers are high. Congested traffic with frequent idling worsens pollution. Poor road conditions and lack of traffic management can spike emissions regardless of vehicle count. Industries, biomass burning, waste incineration, and power generation all contribute significantly to urban pollution but none of which are reflected in per capita vehicle data. Coal-fired and natural gas plants, which burn fossil fuels at high temperatures also emit NOx. During various operations, manufacturing facilities, refineries, and chemical plants emit also emit NO_x. Additionally, biomass burning, forest fires and crop residue burning release NO_x into the atmosphere. Typically, in India, NO_2 levels are noticed as shown in Table 1.3.

Reactions of thermal NO_x occurring at temperatures above 1300 °C in engines and power plants convert inert N_2 to NO and NO_2 as below

$$N_2 + O \rightarrow NO + N \tag{1.5}$$

$$N + O_2 \rightarrow NO + O \tag{1.6}$$

$$N + OH \rightarrow NO + H \tag{1.7}$$

These reactions are temperature-dependent and dominate in gas-fired systems. NO is converted to NO_2 as below

$$NO + O_3 \rightarrow NO_2 + O_2 \tag{1.8}$$

$$NO + HO_2 \rightarrow NO_2 + OH \quad (1.9)$$

$$NO_2 + \text{sunlight} \rightarrow NO + O \rightarrow O_3 \text{ formation} \quad (1.10)$$

NO_2 acts as a precursor to tropospheric ozone and oxidizes into HNO_3, which causes acid rain.

Rapid urban expansion and industrialization have positioned megacities as major contributors to RNS emissions. Cities like Delhi are experiencing rising concentrations of nitrogen dioxide (NO_2), primarily due to fossil fuel combustion from automobiles, heavy industries and transboundary pollution from nearby states in the form of crop residue burning and brick kiln emissions. Elevated NO_2 levels severely degrade urban air quality, posing risks to respiratory health and contributing to atmospheric chemical reactions that result in acid rain. In tandem with nitrate (NO_3^-), these nitrogen species create acidic conditions that harm ecosystems, infrastructure, and water bodies. The exponential rise in vehicular traffic and population density across urban centres has exacerbated nitrogen loading in the environment. Long-term studies reveal a significant increase in the fluxes of ammonium (NH_4^+) and nitrate (NO_3^-) in Delhi's atmosphere and surface systems since 1994, indicating intensifying nitrogen deposition and pollution (Singh et al. 2014).

1.4 Ammonia (NH_3)

Ammonia (NH_3) is another most abundant RNS in the atmosphere with agriculture identified as its primary source (Anderson et al. 2003). Key anthropogenic contributors within agricultural systems include inorganic fertilizer application, livestock manure, enhanced biological nitrogen fixation, biomass burning, and crop residue incorporation following harvest. According to Misselbrook et al. (2000), the agricultural sector is responsible for approximately 90% of global atmospheric NH_3 emissions. As an alkaline gas, ammonia plays a pivotal role in modulating atmospheric acidity. It does so by interacting with acidic constituents such as sulfuric and nitric acids to form ammonium salts (NH_4^+), thereby influencing aerosol chemistry and contributing to neutralization processes (Aneja et al. 2008).

Ammonia is also recognized as a criteria air pollutant under ambient air quality standards (NAAQS). 100 $\mu g/m^3$ and 400 $\mu g/m^3$ NH_3 limits are defined for annual and 24 h exposure, respectively for urban and industrial sites in India (CPCB 2009). Table 1.4 gives NH_3 levels at different sites in South Asia.

Approximately 81% of global NH_3 emissions originate from agriculture, However, urban sectors are increasingly contributing to ammonia pollution via vehicular emissions from catalytic converters, waste treatment processes, sewage systems, and industrial activities such as fertilizer manufacturing and synthetic ammonia production. Ammonia is also released through natural processes, including microbial activity in soil and vegetation decay, volatilization from oceans, and wildfires that

Table 1.4 NH_3 levels at different sites in South Asia

Site	Country	NH_3 ($\mu g\ m^{-3}$)	References
Okhla, Delhi	India	40.7	Singh and Kulshrestha (2014)
Mai village, Uttar Pradesh	India	51.6	Singh and Kulshrestha (2014)
Lahore	Pakistan	50.1	Biswas et al.(2008)
Kanpur	India	22.3	Behera and Sharma (2010)
JNU, Delhi	India	29.4	Singh and Kulshrestha (2012)

combust organic matter. NH_3 plays a crucial role in the formation of fine particulate matter ($PM_{2.5}$), with ammonium-rich particles associated with chronic respiratory illnesses and premature mortality. The complexity of NH_3 sources across both rural and urban landscapes underscores its critical status in air quality management and the urgent need for targeted mitigation strategies.

1.4.1 Urea Hydrolysis an Important Source of NH_3

Urea is broken down by the enzyme urease, naturally present in soil:

$$(CONH_2)_2 + H_2O \xrightarrow{\text{urease}} NH_2COOH \text{ (carbamic acid)} + NH_3 \tag{1.11}$$

Carbamic acid is unstable and rapidly decomposes:

$$NH_2COOH \rightarrow NH_3 + CO_2 \tag{1.12}$$

Net result is that 1 molecule of urea produces 2 molecules of NH_3 and 1 molecule of CO_2.

Ammonia Equilibrium in Aerosol Droplet

NH_3 reacts with water:

$$NH_3 + H_2O \rightleftharpoons NH_4^+ + OH^- \tag{1.13}$$

At high pH of aerosol droplet, equilibrium shifts toward NH_3, increasing volatilization. At low pH, more NH_4^+ forms, reducing NH_3 loss. Due to this reason, NH_4/NH_3 ratios are higher in temperate regions than in Indian region (Singh and Kulshrestha 2012). Also, Singh and Kulshrestha (2012) have attributed the elevated levels of gaseous ammonia (NH_3) in Delhi partly to the presence of alkaline dust particles in the atmosphere. Alkaline soil dust, rich in compounds like calcium carbonate, creates a favourable environment for NH_3 to remain in the gas phase. Normally, NH_3 would react with acidic species (like H_2SO_4 or HNO_3) to form particulate ammonium salts (NH_4NO_3, $(NH_4)_2SO_4$).But in Delhi, the high concentration of alkaline dust neutralizes these acids, reducing the conversion of NH_3 into particulate form, and thus keeping NH_3 levels elevated. NH_3 is a weak base, and in the presence of alkaline aerosols, its tendency to form ammonium particles is suppressed. This leads to a higher fraction of NH_3 in the gaseous phase, especially in regions with frequent construction activity, dry conditions, and exposed soil, all common in urban Delhi.

1.5 NH_3 Emissions During Different Stages of Cultivation and Crop Residue Burning

Fertilizer application is recognized as a major contributor to atmospheric ammonia (NH_3) emissions, which significantly affect both air quality and the nitrogen biogeochemical cycle. To better understand the temporal dynamics of NH_3 release, Yadav et al. (2023) conducted a study which monitored diurnal and seasonal variations in NH_3 concentrations across various crop stages during the Kharif and Rabi seasons, from July 2017 to June 2018.The findings revealed that NH_3 levels were markedly lower following the basal dressing of DAP (diammonium phosphate) compared to the top dressing of urea, highlighting the influence of fertilizer type and application method on emission intensity.

As shown in Fig. 1.5, NH_3 emissions were much higher during crop residue burning period in comparison to the post fertilizing period. Hence, controlling the crop residue burning will also help in controlling NH_3 in air.

Seasonal Trends in NH_3 Concentrations:

Kharif Season NH_3 concentrations varied during *sowing*: 1.4–45.2 $\mu g\ m^{-3}$, *fertilizer addition*: 63.1–190.9 μgm^{-3}, *grain filling*: 98.9–187.5 μgm^{-3}, *crop residue*

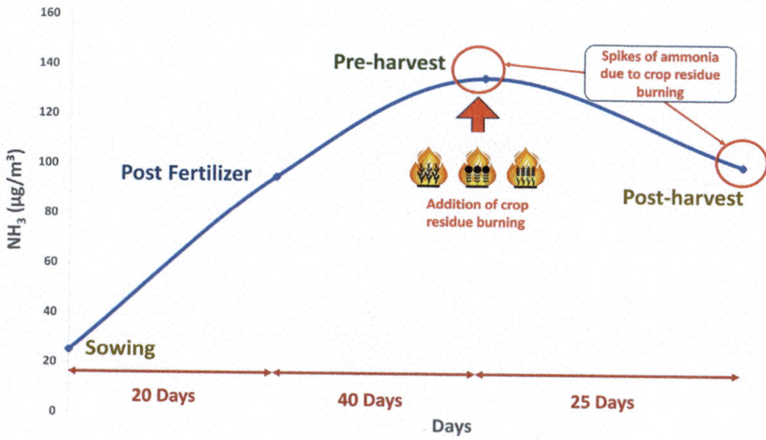

Fig. 1.5 NH_3 measurements during different steps of cultivation and at the time of crop residue burning in Haryana state during 2017–18 (Source data: Yadav et al. 2023)

burning: 56.8–249.5 μgm^{-3}, with an average NH_3 concentration: 125.3 μgm^{-3}. *Rabi* Season NH_3 varied during *sowing*: 22.9–68.4 μgm^{-3}, *Crown Root Initiation (CRI)*: 59.4–104.71 μgm^{-3}, *Panicle initiation*: 26.3–56.0 μgm^{-3}, *grain filling*: 48.2–147.2 μgm^{-3}, *maturity*: 21.5–80.4 μgm^{-3} with an average NH_3 concentration: 51.8 μgm^{-3} (Yadav et al. 2023).

The data clearly indicates that Kharif season exhibited significantly higher NH_3 emissions than Rabi, likely due to climatic factors and fertilizer practices. Furthermore, a noticeable dip in NH_3 levels was observed in the transition between the two cropping seasons. This decline can be attributed to monsoonal rainfall, which facilitates atmospheric scavenging of ammonia, and gas-to-particle conversion, which reduces the gaseous NH_3 fraction.

The increased production and use of nitrogen fertilizers contribute to elevated levels of reactive nitrogen species (RNS) in the environment, including nitrous oxide (N_2O), nitrate (NO_3^-), and ammonium (NH_4^+). These compounds not only impact soil health and water quality but also drive climate change, with N_2O being a potent greenhouse gas. More details are given in Chap. 4.

1.6 N Air Pollution Hotspots in South Asia

Fine-mode ammonium aerosols which are generally in the form of NH_4NO_3 and $(NH_4)_2SO_4$ are effectively scavenged by rainfall, influencing not only air quality and rainwater chemistry, but also impacting soils, vegetation, and urban infrastructure through wet deposition. Based on the NO_3 and NH_4 concentrations in precipitation, air pollution hotspots have been identified in South Asia (Fig. 1.6). The Indo-Gangetic Plain and Western Bangladesh have emerged as significant hotspots for reactive nitrogen, particularly in the form of ammonium (NH_4^+) and nitrate (NO_3^-). These elevated concentrations are strongly linked to intense agricultural activity, vehicular emissions, and urbanization in the region. Additionally, the Indo-Gangetic Plain also shows notably high levels of sulfate (SO_4^{2-}), indicating substantial contributions from industrial and fossil fuel combustion sources. (Kulshrestha et al. 2024). Several monitoring sites—including GAU (Bangladesh), DTG (Sri Lanka), BSC (Bhutan), and Pokhara (Nepal) have recorded a concerning frequency of acidic rain events, with more than one-third of rainfall episodes exhibiting pH below 5.6. These findings signal potential environmental stress on regional ecosystems, infrastructure, and public health. This research was conducted under the UKRI GCRF South Asian Nitrogen Hub led by the Centre for Ecology and Hydrology, Edinburgh, UK, presents findings on rainwater pH and chemical composition across multiple South Asian locations during 2022–2023.

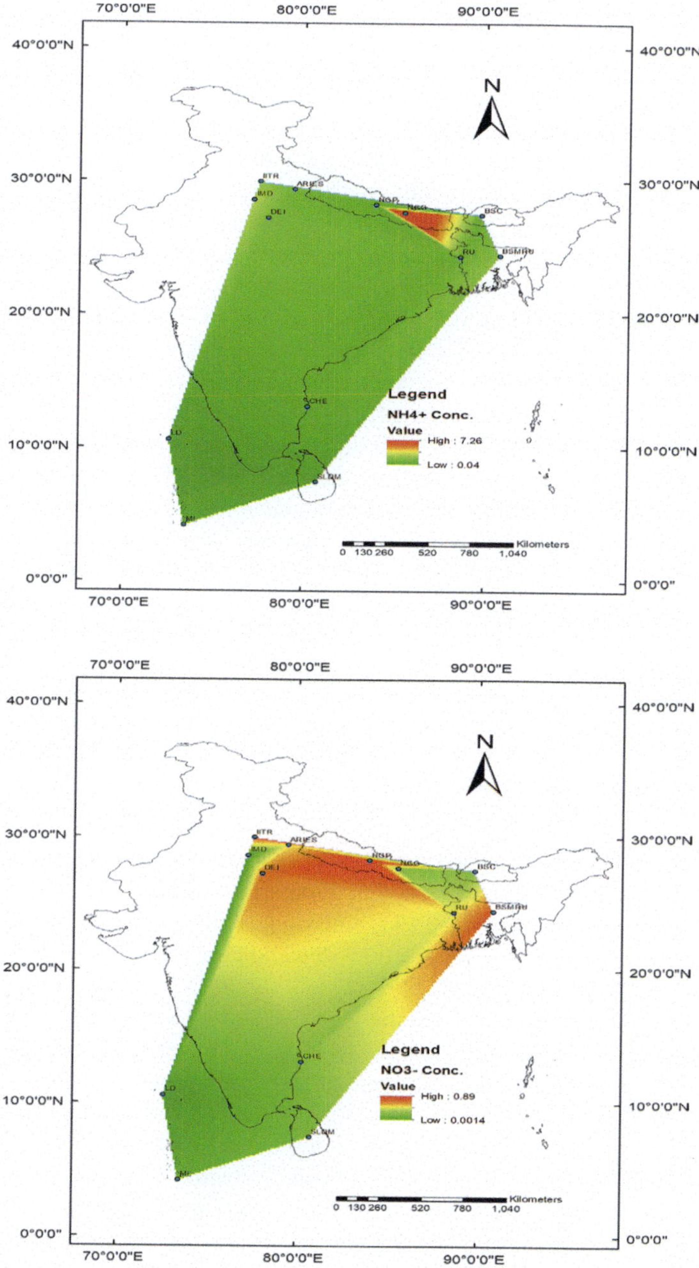

Fig. 1.6 Hot spots and spatial distribution of NO_3 and NH_4 in rain water across South Asian Nitrogen Hub (SANH) in South Asia (mgL^{-1}) (*Source* Kulshrestha et al. 2024)

1.7 Indoor Nitrogen Pollution

Human actions play a significant role in determining indoor air quality. In homes that rely on natural ventilation, indoor nitrogen pollution is controlled not only by everyday activities like cooking and cleaning but also by outdoor influences such as industrial emissions and traffic pollution. This interplay highlights the urgent need for comprehensive strategies that integrate both indoor and ambient RNS management to ensure healthier living conditions in urban settings. A study by Katoch and Kulshrestha (2021) revealed that the concentration of trace gases followed a consistent pattern at two urban households-DT (Dwarka Household) and MH (Mayapuri Household) in Delhi. Ammonia (NH_3) was most abundant, followed by nitrogen dioxide (NO_2). NH_3 accounted for approximately 90% of reactive nitrogen (Nr) species at the DH site and 85% at the MH site, with peak levels during the monsoon season, while NO_2 concentrations were higher in winter. Oxidation ratios, used to assess gas-to-particle conversions, indicated the sequence $NO_2 > NH_3$. Source apportionment techniques, including principal component analysis and mass ratio evaluation, revealed distinct pollution origins at each site. At the DH household, indoor sources such as cooking, cleaning, biomass burning, and dust resuspension played key roles in shaping particulate composition. In contrast, the MH site was predominantly influenced by outdoor factors such as coal combustion and emissions from nearby industrial activities.

1.8 NH_3 as an Alternate Fuel

NH_3 is a very useful compound in energy world. As the world moves toward achieving carbon neutrality by 2050, low-carbon ammonia is emerging as a next-generation fuel with strong potential to reduce CO_2 emissions, in the beginning as the maritime fuel. Ammonia is poised to become a high-demand clean energy source for the next generation.

Green ammonia is produced via the Haber–Bosch process, but in this process, Hydrogen is generated through electrolysis of water, powered by renewable energy (solar, wind, hydro), and the nitrogen is extracted from the air. When it is produced using hydrogen derived from natural gas, with carbon capture and storage (CCS) to reduce emissions, it is termed as blue ammonia. The global green ammonia market is witnessing explosive growth, fueled by its promise as a clean energy solution and a zero-carbon fuel alternative. Valued for its sustainability and versatility, the market is projected to soar to $6.16 billion by 2030, expanding at a remarkable CAGR (Compound Annual Growth rate) of 66% between 2024 and 2030 (https://www.marketsandmarkets.com/Market-Reports/green-ammonia-market-118396942.html). Innovations in ammonia combustion and fuel cell technologies are actively tackling challenges related to toxicity and low energy density, with the goal of enhancing both safety and performance.

Chapter 2
N Deposition

2.1 Evidences of Increase of RNS in South Asia

In a study on snowfall chemistry at Gulmarg, data from 2012–2013 was compared with 1986–87 (Kumar et al. 2016). In this study it was found that over nearly three decades, NO_3^- and NH_4^+ showed almost doubling of concentrations. NO_3^- increase was linked to vehicular emissions, especially from diesel generators and rising tourist traffic, part from their long range transport. NH_4^+ levels rose due to human and animal waste, notably from an increase in horses used for tourism and poor sanitation practices. Average pH dropped from 6.7 (1986–1987) to 5.9, indicating increased acidity. Elevated $nssSO_4^{2-}/nssCa^{2+}$ and $NO_3^-/nssCa^{2+}$ ratios suggest reduced buffering capacity and stronger acid inputs. The shift in snowmelt composition signals potential risks to soil health, water quality, and ecosystem stability. Figure 2.1 shows the location of Gulmarg.

In another study by Singh et al. (2017), a dramatic rise in reactive nitrogen (Nr) deposition over Delhi between 1994 and 2013 has been reported. Nitrate (NO_3^-) concentrations in rainwater surged by 486%, while ammonium (NH_4^+) levels rose by 283%, driven by intensified vehicular emissions, human activities, and changes in the city's land use and land cover (LULC).

2.2 Acid Rain

Acid rain refers to any form of precipitation—rain, snow, fog, or even dust that contains acidic components like sulfuric acid (H_2SO_4) and nitric acid (HNO_3). Among various RNS, NO and NO_2 are oxidized forming HNO_3 giving rise to acidic precipitation.

The dominating process of NO_2 oxidation during day time is through OH-

U. C. Kulshrestha et al., *Atmosphere-Biosphere Interactions of Reactive Nitrogen in South Asia*, Springer Briefs in Interdisciplinary Geosciences – Asia-Pacific,
https://doi.org/10.1007/978-3-032-17194-8_2

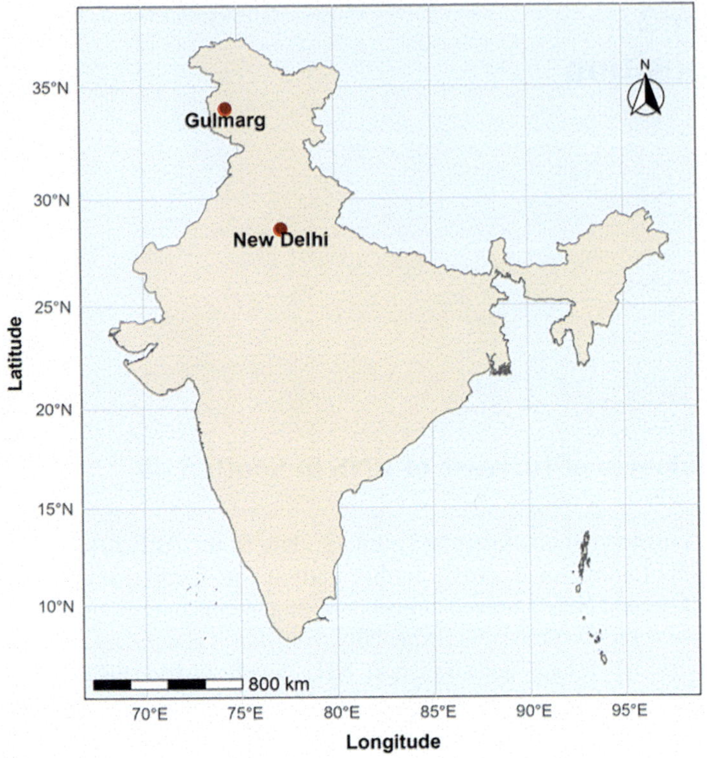

Fig. 2.1 Location of Gulmarg and New Delhi

$$NO_2 + OH \rightarrow HNO_3 \tag{2.1}$$

Further, NO_2 reacts with O_3 in the afternoon forming NO_3 which further reacts with NO_2 forming N_2O_5:

$$NO_2 + O_3 \rightarrow NO_3 + O_2 \tag{2.2}$$

$$NO_3 + NO_2 \rightarrow N_2O_5 \tag{2.3}$$

Aqueous-phase reaction of N_2O_5 during nighttime largely contributes to HNO_3 formation causing the acidification-

$$N_2O_5 + H_2O \rightarrow 2HNO_3 \tag{2.4}$$

The nitric acid formed is highly soluble in water and gets incorporated into rain, snow, or fog lowering the pH of precipitation. Table 2.1 gives NO_3 concentrations

Table 2.1 pH and NO_3 content in rain water at different sites in South Asia (Kulshrestha et al. 2024)

Site	Country	pH	NO_3 (mg NL^{-1})
Gazipur	Bangladesh	5.46	1.29
Rajshahi	Bangladesh	6.41	0.34
Male	Maldives	6.48	17.45
Peradeniya	Sri Lanka	5.60	0.19
Kanglung,	Bhutan	5.80	0.21
Pokhara	Nepal	5.91	0.73
Kathmandu	Nepal	6.45	0.25
Chennai	India	6.04	0.23
Lakshadweep	India	6.05	0.03
Agra	India	6.47	0.54
Roorkee	India	7.18	0.73
New Delhi	India	7.51	0.05
Nainital	India	5.73	0.45

in rain water at different pH in South Asia. Figure 2.2 shows the locations of these sites.

Asia ranks third globally in nitrogen deposition rates, trailing behind North America and Europe. This positioning is backed by multi-model analyses that estimate anthropogenic reactive nitrogen (Nr) budgets across continents, highlighting the substantial impact of industrialized regions. In India, nitrogen use in agriculture has soared dramatically. Nitrogenous fertilizer consumption skyrocketed from just 1 million tonnes in the 1960s to over 15 million tonnes by 2010–11 a nearly 15- fold increase. This surge is closely tied to intensifying agricultural demands and food security policies. The Indo-Gangetic Plain, a key agricultural zone, has emerged as a hotspot for nitrogen abundance due to concentrated fertilizer application and dense population activity. A detailed and authoritative Nitrogen Assessment for India was undertaken by Abrol and colleagues in 2017, offering a comprehensive evaluation of sources, impacts, and mitigation strategies related to nitrogen management in the country.

2.3 Eutrophication

HNO_3 is deposited via rain (wet deposition) or as gases/particles (dry deposition). Once in the soil or water, it is dissociated into nitrate-

$$HNO_3 \rightarrow NO_3^- + H^+ \tag{2.5}$$

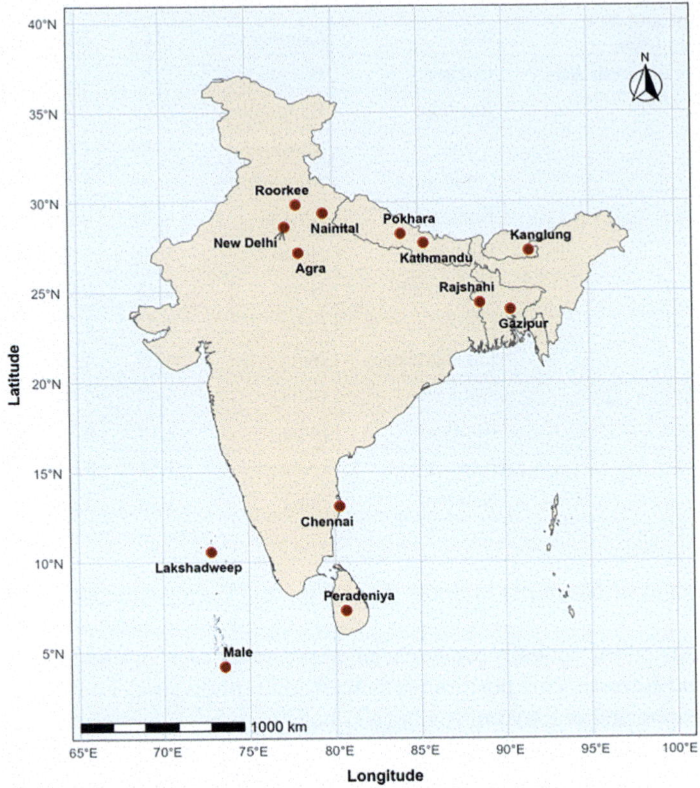

Fig. 2.2 Location of sites of pH and NO_3 in rain water in South Asia

NO_3^- acts as nutrients, fueling excessive growth of algae and aquatic plants known as eutrophication. Accelerated nutrient loading from NO_2-derived nitrates intensifies eutrophication. When algae die, microbial decomposition consumes dissolved oxygen leading to hypoxia, harming fish and other aquatic life

$$C_6H_{12}O_6 + 6O_2 \rightarrow 6CO_2 + 6H_2O \tag{2.6}$$

Algal blooms block sunlight, disrupt aquatic food chains, and reduce biodiversity. Decomposition of excess biomass lowers oxygen levels, creating dead zones.

NO_2 and NO_3 from atmospheric sources including automobile emissions and fertilizer runoff are key drivers of nutrient enrichment in water bodies. Denitrification and microbial processes attempt to restore balance, but excess nitrogen overwhelms natural systems.

NH_3 also results in eutrophication. NH_3 dissolves in water to form ammonium (NH_4^+), a readily assimilable nitrogen source for phytoplankton and aquatic plants

$$NH_3 + H_2O \rightarrow NH_4^+ + OH^- \tag{2.7}$$

Table 2.2 Examples of eutrophication in South Asia

Waterbody	City	Country	References
Dal lake	Kashmir	India	https://organicabiotech.com/causes-of-pollution-in-dal-lake/
Kolleru lake	Andhrapradesh	India	https://www.google.com/search?client=firefox-b-d&q=Kolleru+lake
Nazafgarh lake	Delhi	India	https://www.unep.org/news-and-stories/story/how-reduce-pollution-delhis-waterways-study
Dhanmondi lake	Dhaka	Bangladesh	https://doi.org/10.1007/s10661-024-13419-y
Gulshan lake	Dhaka	Bangladesh	https://doi.org/10.1007/s10661-024-13419-y
Banami lake	Dhaka	Bangladesh	https://doi.org/10.1007/s10661-024-13419-y

This boosts primary productivity, often tipping the nutrient balance toward algal blooms reducing light penetration which disrupts aquatic food webs, and alter biogeochemical cycling.

Through nitrification cascade, NH_4^+ is oxidized to nitrite (NO_2^-) and then nitrate (NO_3^-) by nitrifying bacteria

$$NH_4^+ \rightarrow \text{Nitrosomonas} \rightarrow NO_2^- \rightarrow \text{Nitrobacter} \rightarrow NO_3^- \quad (2.8)$$

These transfromations further increase nitrate levels, perpetuating eutrophication. Table 2.2 gives some examples of eutrophication in South Asia. Figure 2.3 shows the location of these waterbodies.

2.4 Formation of NH_4NO_3 Aerosols in the Indo-Gangetic Region

During the ISRO Land Campaign-II in winter 2004, at Allahabad site, Kulshrestha et al. (2010) found that a significant portion of the water-soluble fraction of PM_{10} aerosols consisted primarily of ammonium nitrate (NH_4NO_3) and various forms of ammonium sulfate, including $(NH_4)_2SO_4$, NH_4HSO_4, and $(NH_4)_3H(SO_4)_2$. The concentrations of sulfate (SO_4^{2-}), nitrate (NO_3^-), and ammonium (NH_4^+) aerosols were observed to rise markedly with increasing relative humidity, particularly up to their deliquescence point (~63% RH). This trend suggests that elevated humidity levels enhance the transformation and partitioning of the gaseous precursors sulfur dioxide (SO_2), nitrogen oxides (NO_x), and ammonia (NH_3) into their particulate forms. Furthermore, as ambient temperatures declined over the course of winter, concentrations of NO_3^- and NH_4^+ increased, likely driven by the semi-volatile nature of ammonium nitrate, which favors aerosol stability under cooler conditions.

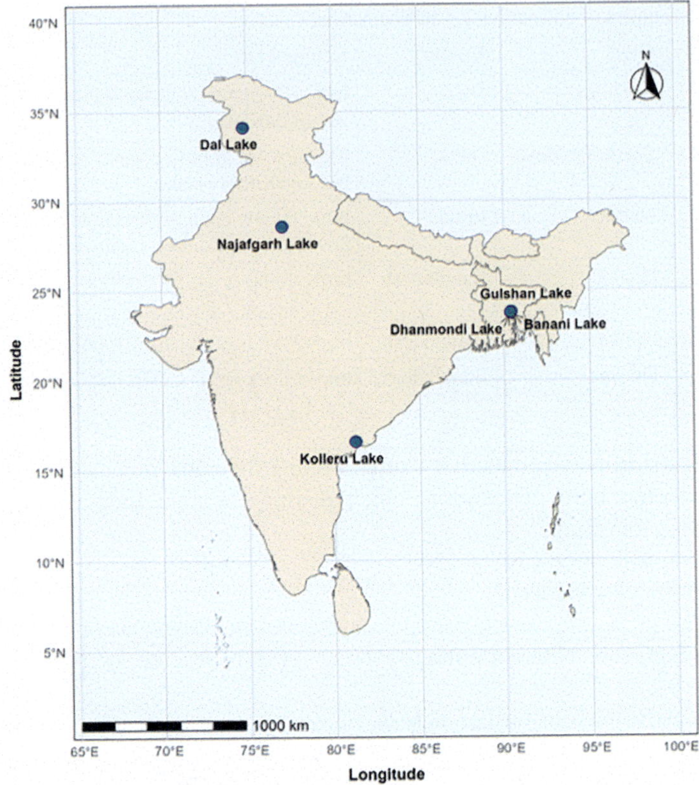

Fig. 2.3 Location of water bodies with eutrophication

$$NH_3(g) + HNO_3(g) \leftrightarrow NH_4NO_3(p) \tag{2.9}$$

here g: gas and p: particle

A possibility of NH_4NO_3 aerosol formation across the Delhi region due to favorable thermodynamic conditions (Tiwari and Kulshrestha 2019). This study investigated the spatial and diurnal behavior of inorganic RNS NH_3, NO_x, HNO_3 and their particulate counterparts NH_4^+, NO_3^- across three urban sites in India's National Capital Region (Faridabad, Delhi, and Rohtak) with distinct land use profiles. The study identified Faridabad as the most polluted site and highlights NH_3 as the dominant RNS component at all locations. This study reveals a clear spatial gradient in RNS gases across urban NCR sites, shaped by land use and local emissions, with HNO_3 showing strong site-specific variability and NH_3 dominating total nitrogen fractions due to urban-agricultural interactions. Lower temperature ranges were found to be favourable for NH_4NO_3 aerosol formation.

Table 2.3 AAI values at some sites in South Asia

Sampling site	Country	AAI
Gazipur	Bangladesh	74.48
Male	Maldives	472.90
Kanglung	Bhutan	102.71
Chennai	India	66.19
Pokhara	Nepal	787.32
Kathmandu	Nepal	168.67
Peradeniya	Sri Lanka	458.70

2.5 Ammonium Availability Index (AAI)

To assess the ability of ammonium ions (NH_4^+) to neutralize nitrate (NO_3^-) and sulfate (SO_4^{2-}) in precipitation, the Ammonium Availability Index (AAI) is a good indicator which establishes the molar balance between NH_4^+ concentration and the quantity needed to fully neutralize sulfuric and nitric acids. Table 2.3 gives AAI values at some sites in South Asia (Kulshrestha et al. 2023). If AAI < 100%, it indicates an ammonium deficit, meaning sulfate and nitrate are present in excess. An AAI = 100% suggests complete neutralization, where SO_4^{2-} and NO_3^- exist as $(NH_4)_2SO_4$ and NH_4NO_3, respectively (Szep et al. 2018). If AAI > 100%, sufficient ammonium is available to fully neutralize both acid ions. It's important to note that AAI reflects only the neutralization potential related to sulfate and nitrate, and it does not directly correlate with the rainwater pH, as other cations like Ca^{2+}, Mg^{2+}, K^+, and Na^+ also influence pH (Kulshrestha et al. 1996). Most of the sites in South Asia show AAI values more than 100% indicating sufficient NH_4 to neutralize the acidity of (NO_3^-) and sulfate (SO_4^{2-}). However, one site in Bangladesh and one site in India show AAI values below 100%, indicating insufficient ammonium for full neutralization of acidity. Figure 2.4 shows the locations of AAI sites.

2.6 N Deposition Budgets

South Asia experiences both wet and dry deposition of nitrogen compounds. Wet N deposition is more prominent during the monsoon season, especially in the Indo-Gangetic Plain and Western Bangladesh, due to high population density and agricultural activities in these pockets. Dry deposition occurs year-round and is influenced by dust, aerosols, and gaseous Nr species. N deposition in the form of coarse particles through dustfall is also significant in this region due to high loading of mineral dust particles in the atmosphere (Kulshrestha 2017).

Most of the sites show high levels of wet nitrogen deposition, especially in the form of ammonium (NH_4^+–N) (Kulshrestha et al. 2005). In general, urban and industrial sites show higher NO_3^-–N deposition, while rural and agricultural sites have elevated

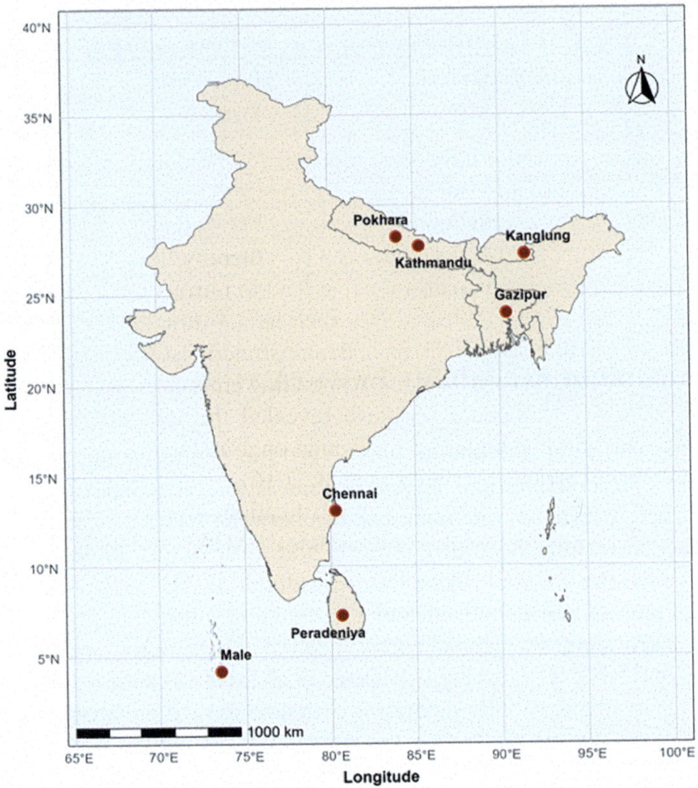

Fig. 2.4 Locations of AAI sites in South Asia

NH_4^+–N levels. However, some urban sites show higher NH_4^+ wet deposition due to localized sources of NH_3. The higher NH_4^+/NO_3^- ratio indicates a strong influence from agricultural sources, such as fertilizer use and animal waste (Kulshrestha et al. 2024). Total Inorganic N (TIN) which comprises of ($NH_4^+ + NO_3^-NO_3^-$) is given in Table 2.4 and sites are shown in Fig. 2.5.

Compared to monitoring sites across South Asia, the remote Japanese EANET sites exhibit a lower magnitude of wet nitrogen deposition. This pattern suggests a comparatively weaker regional emission intensity or source strength influencing these Japanese locations. These sites are shown Fig. 2.5. The lower deposition levels are consistent with the geographic locations and lower anthropogenic pressure at these sites, particularly those situated on islands or in high-altitude forested regions. These values also help in understanding temperate vs tropical wet deposition of NH_4^+ and NO_3^-.

Nevertheless, a closer examination of the long-term wet deposition data (as summarized in Table 2.4) reveals that at several Japanese sites specifically Rishiri, Oki, Hedo, and Ogasawara the concentration of ammonium nitrogen (NH_4^+–N) in

Table 2.4 Total inorganic N (TIN) in wet deposition at different sites in India (kg N ha^{-1} yr^{-1})

Sites	NH_4^+–N/ NO_3^-–N	N–NH_4^+	N–NO_3^-	TIN	References
Delhi, India	5.19	2.7	0.52	3.22	Kulshrestha et al. (1996)
Hudegadde, India	6.91	6.22	0.90	7.12	Kulshrestha et al. (2014)
Hyderabad, India	0.82	11.66	14.22	25.88	Kulshrestha et al. (2014)
Agra, India	25.92	23.33	0.90	24.23	Saxena et al. (1991)
Pune, India	0.49	1.33	2.69	4.02	Rao et al. (2014)
Bhosari, India	0.30	2.99	9.70	12.69	Rao et al. (2014)
Delhi, India	3.42	10.45	3.05	13.5	Singh et al. (2017)
Jaunpur, India	2.04	3.19	1.56	4.75	Singh et al. (2017)
Rishri, Japan	1.25	2.5	2.0	4.5	Ban et al. (2016)
Tappi, Japan	0.86	2.6	3.0	5.6	Ban et al. (2016)
Sado-Saki, Japan	0.84	2.7	3.2	5.9	Ban et al. (2016)
Oki, Japan	1.0	3.1	3.1	6.2	Ban et al. (2016)
Happo, Japan	0.75	2.8	3.7	6.5	Ban et al. (2016)
Yushuhara, Japan	0.89	2.6	2.9	5.5	Ban et al. (2016)
Hedo, Japan	1.34	3.1	2.3	5.4	Ban et al. (2016)
Ogasawara, Japan	1.37	1.1	0.8	1.9	Ban et al. (2016)

precipitation consistently exceeds that of nitrate nitrogen (NO_3^-–N). This dominance of reduced nitrogen species in wet deposition suggests a notable influence of regional or long-range transported ammonia emissions, potentially from agricultural or marine sources, even in these relatively pristine environments. It seems that the regional emissions sources of NH_3 likely contributed to an overall higher side total nitrogen deposition across Japan.

2.7 NH_3 Deposition Leads to Soil Acidification

Ammonia itself is not acidic, but it reacts with acidic compounds in the atmosphere. It combines with nitric acid (HNO_3) and sulfuric acid (H_2SO_4) formed from NO_x and SO_2 emissions to create ammonium salts like NH_4NO_3 and $(NH_4)_2SO_4$. These salts are part of particulate matter ($PM_{2.5}$) and can be deposited via precipitation, contributing to acidic deposition. While ammonium salts may temporarily neutralize acidity, their deposition still leads to soil acidification over time. Once deposited, ammonium ions (NH_4^+) undergo microbial oxidation in soil, releasing hydrogen ions (H^+):

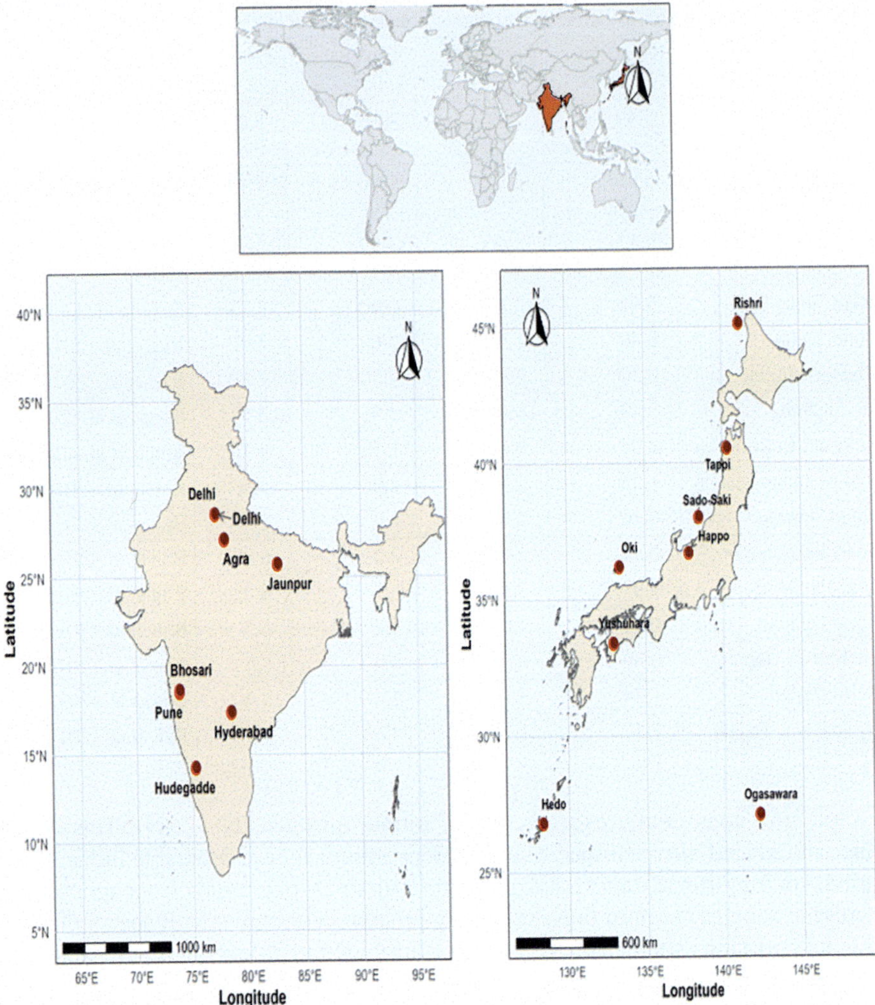

Fig. 2.5 Sites for which NH_4 and NO_3 fluxes are reported

1. Step 1: Ammonium to Nitrite

$$NH_4^+ + 1.5O_2 \rightarrow NO_2^- + 2H^+ + H_2O \tag{2.10}$$

2. Step 2: Nitrite to Nitrate

$$NO_2^- + 0.5O_2 \rightarrow NO_3^- \tag{2.11}$$

These reactions lower soil pH, leading to acidification, especially in ecosystems with low buffering capacity. Regions where acid rain occurrence is still a common phenomenon can be considered as low buffering capacity areas. These areas show low Ca and Mg content in precipitation and soils.

Excessive or imbalanced N fertilization is a major problem in intensive land use systems in different countries since it increases soil degradation, nutrient losses and environmental contamination (Rahman, 2014; Islam et al. 2018; Rahman et al. 2021; Hasnat et al. 2022).The global arable land is declining continuously, while in Bangladesh annual land loss is about 87,000 ha (Hasan et al. 2019b; Rahman et al. 2021). Therefore, a balanced supply of N fertilizer along with other essential nutrients is needed for optimum yields of different crops from limited land resources (Shashi et al. 2018; Alam et al. 2023; Anik et al. 2023).

Urea is the most widely used nitrogen-based fertilizer. It is produced through the Haber–Bosch process a method that converts inert atmospheric nitrogen (N_2) into ammonia (NH_3) under high temperature and pressure using hydrogen derived primarily from natural gas. This ammonia is then further processed to produce urea. Crucially, for every molecule of NH_3 generated, one molecule of carbon dioxide (CO_2) is released, making fertilizer manufacturing a significant source of greenhouse gas emissions.

Since the Green Revolution, urea consumption in India has surged from 1 million tons to approximately 33.5 million tons. In Bangladesh, urea consumption in agriculture during 1960–1965 was about 68,000 tons only, while it increased to 2.59 million tons during 2015–2019. Increased fertilizer production has driven higher RNS emissions, with urea being a primary contributor. Manufactured through the Haber–Bosch process, urea production transforms inert nitrogen (N_2) into NH_3, with each molecule of NH_3 generation releasing an equivalent molecule of CO_2. Additionally, the transportation sector emits both NO_x and CO_2, further linking the reactive nitrogen problem to global warming and climate change. Fertilizer application in soils also releases N_2O, the strongest greenhouse gas contributing to climate change.

Haber–Bosch Process involves atmospheric nitrogen and hydrogen (typically derived from natural gas) under high temperature and pressure, with iron catalyst.

Step 1

Generation of H_2, which also releases CO_2

$$CH_4 + H_2O \rightarrow CO + 3H_2 \tag{2.12}$$

$$CO + H_2O \rightarrow CO_2 + H_2 \tag{2.13}$$

And then use of H_2 and N_2

$$N_2 + 3H_2 \rightarrow 2NH_3 \tag{2.14}$$

Step 2

Ammonia produced in Step 1 reacts with carbon dioxide to form urea:

$$2NH_3 + CO_2 \rightarrow NH_2CONH_2 + H_2O \tag{2.15}$$

2.8 Long Range Transport of RNS Over Himalayan Sites

Long-range transport of pollutants can have adverse impacts on the environment of a location or region. Sometimes distant pollution sources are more harmful than the local pollution sources. Gaseous species are generally transported to very far places through airmasses. While travelling, these species are oxidized and deposited generally via precipitation. The findings of a snow chemistry study within the Himalayan region during winter seasons spanning 2016 to 2019 revealed a substantial spatial variability in dissolved inorganic nitrogen $\left(DIN = NH_4^+ + NO_3^-\right)$, with concentrations peaking at 75 μmol L^{-1} in the rural site (Nirmand) and decreasing to 38 μmol L^{-1} at the urban site (Leh). Moderate at semi-urban site (Beerwah). Locations of these three sites are shown in Fig. 2.6. In contrast, dissolved organic nitrogen levels were found to be highest in the suburban zone and lowest in urban areas. Inorganic ammonium consistently dominated total nitrogen (TN) fractions across all locations, while DON made a notable contribution to TN, ranging from 7 to 14% (Sharma and Kulshrestha 2022). Enrichment factor and principal component analyses indicated that nitrogen wet deposition particularly NH_4^+ and NO_3^- were largely influenced by anthropogenic emissions. Backward trajectory modeling further suggested that air masses contributing to nitrogen deposition at the Leh site originated predominantly from northwestern Europe, while Beerwah received pollution primarily from Afghanistan (Fig. 2.7). The Nirmand site was affected by both long-range transport from Afghanistan and regional sources within India, including Rajasthan, Punjab, and Haryana. These findings are vital for understanding nitrogen deposition dynamics and serve as a scientific foundation for crafting effective environmental and regulatory policies aimed at mitigating nitrogen pollution in the sensitive Himalayan ecosystems.

2.9 RNS Deposition in Western Ghats Versus Other Sites of Different Characteristics

Reactive nitrogen deposition over Western Ghats site shows very interesting patterns as reported by Kulshrestha et al. (2014). In this study, workers have evaluated patterns and processes related to nitrogen deposition across India focusing on the increasing trends of wet deposition fluxes of nitrate (NO_2^-) and ammonium (NH_4^+) at rural (Hudegadde) and urban (Hyderabad) sites (Fig. 2.8). The Hudegadde site is located

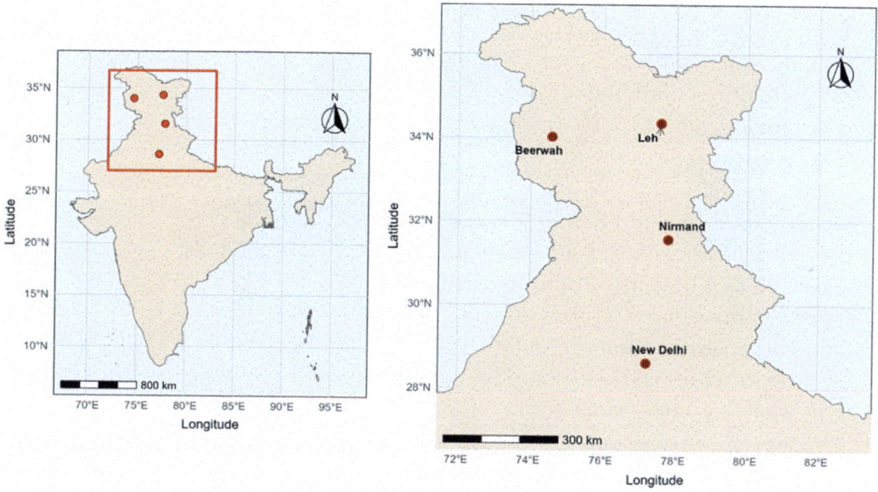

Fig. 2.6 Locations of LRT sites Beerwah, Leh and Nirmand

in Western Ghats biosphere reserve. In this study, NH_4^+ showed a rise at Hudegadde which is likely linked to biomass burning and vegetation decay, while NO_3^- levels escalated in Hyderabad due to urban expansion, vehicular activity, and industrial growth. The research also synthesizes data from multiple studies, classifying sites into five categories (rural, suburban, urban, industrial, and rural-suburban), revealing that urban areas exhibit the highest NO_2^- deposition, whereas industrial zones register peak NH_4^+ levels. Moreover, the Indo-Gangetic region shows strong NH_4^+ fluxes due to ammonia-rich sources. Notably, the study underscores the importance of dry deposition which often exceeding wet deposition levels, as shown for cities like Agra and Delhi yet identifies a significant data gap in dry deposition monitoring. Collectively, these findings stress the urgency for long-term, high-quality nitrogen deposition studies in India to better understand spatial variations, source contributions, and ecological impacts.

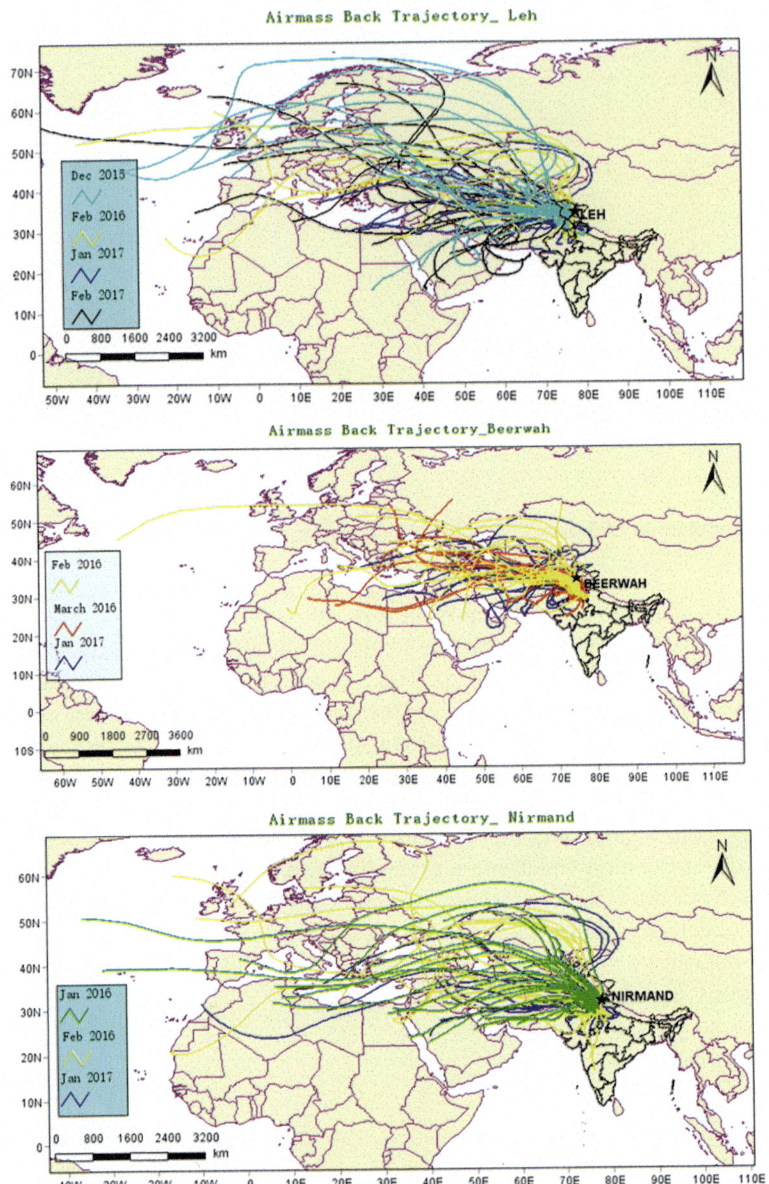

Fig. 2.7 Airmass back trajectory showing long range transport of pollution to Leh, Beerwah and Nirmansites (*Source* Sharma and Kulshrestha 2019)

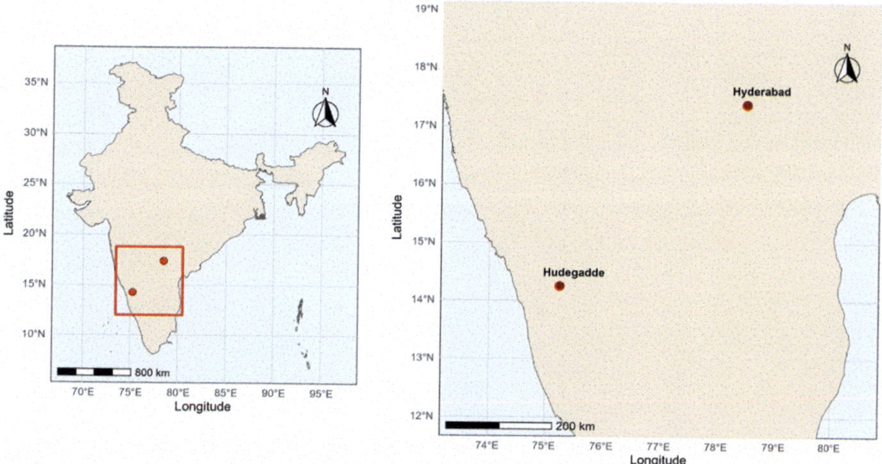

Fig. 2.8 Location of Hyderabad and Western Ghats site (Hudegadde) and Hyderabad in India

Chapter 3
N_2O and Other Greenhouse Gas Concerns

3.1 Agriculture and Non-carbon Dioxide Greenhouse Gases (GHGs)

The evolving climate landscape calls for a broader approach that focuses not only on carbon dioxide (CO_2), but includes consideration of more potent greenhouse gases (GHGs) such as methane (CH_4), nitrous oxide (N_2O), ozone (O_3), and water vapor (H_2O). Global climate policies must therefore integrate the abatement costs of strategies targeting these non-CO_2 gases, while also considering the net radiative forcing from both anthropogenic and agricultural sources (Reay et al. 2012). Among human-driven activities, agriculture stands out as a major source of GHG emissions, contributing significantly to the atmospheric burden of CH_4 and N_2O—two of the most impactful non-CO_2 gases in terms of global warming potential (Bhatia et al. 2010a). The growing urgency around climate change is prompting sustained pressure on both developed and developing nations to take meaningful action in curbing these emissions. For reference, the atmospheric lifetimes and global warming potentials (GWP) of common greenhouse gases are provided in Table 3.1.

Ozone levels are also increasing in the atmosphere and their concentrations may reach above 40 ppb on many days. Such levels are considered to be phototoxic to plants (Singh et al. 2013; Tomer et al. 2015) and play a role in the reactive N cycle and in generating N_2O emissions. The severity of ozone damage to plant systems depends on factors such as ozone concentration, exposure duration, environmental conditions (e.g., sunlight, humidity, temperature), soil water availability, plant genetics and their developmental stage. Plants possess antioxidant defence systems, but prolonged or high ozone exposure can overwhelm these mechanisms (Bhatia et al. 2012).

U. C. Kulshrestha et al., *Atmosphere-Biosphere Interactions of Reactive Nitrogen in South Asia*, Springer Briefs in Interdisciplinary Geosciences – Asia-Pacific,
https://doi.org/10.1007/978-3-032-17194-8_3

Table 3.1 Lifetime and global warming potential (GWP) of CO_2, CH_4 and N_2O as per IPCC AR5 (2014)

Greenhouse gas	Lifetime (years)	Global warming potential (time horizon)		
		20 years	100 years	500 years
CO_2	§	1	1	1
CH_4	12±3	72	25	7.6
N_2O	120±6	264	265	153

(*Source* https://unfccc.int/process/transparency-and-reporting/greenhouse-gasdata/greenhouse-gas-data-unfccc/global-warming-potentials)

3.2 Fertilizer Use in Global Agriculture

Global nitrogen (N) fertilizer use is a critical aspect of modern agriculture that heavily influences crop yields and food security. About half of the world's population today is reliant on synthetic fertilizers for food production. Cereals such as wheat, rice, and maize account for the largest share of global N fertilizer consumption with each accounting for 15 to 18% of the world's total N fertilizer use (Ritchie et al. 2022). China, India, and the United States collectively dominate global N fertilizer consumption, holding an aggregate market share of more than 55% (FAOSTAT 2023). Fruit and vegetables together account for nearly 15% of global N fertilizer use. The use efficiency of applied N fertilizer varies from 30 to 50% depending on the climate and management conditions (Mboyerwa et al. 2022).

3.2.1 Rice and Fertilizer Use

Rice (*Oryza sativa*) is a vital staple food for nearly half of the global population and sustains the livelihoods of approximately 145 million households. Rice fields are also significant sources of greenhouse gases—responsible for around 12% of global methane (CH_4) and 30% of nitrous oxide (N_2O) emissions (Dieleman et al. 2022; Mboyerwa et al. 2022). Due to declining crop responsiveness to fertilizers, farmers often exceed recommended application rates (Bhatia et al. 2010), especially in intensive rice–wheat systems such as those found in the Indo-Gangetic Plain (IGP), which faces a net nutrient deficit of 2.22 million tonnes of NPK annually. This trend contributes to increasing reactive nitrogen losses (Bhatt et al. 2016; Bhatia et al. 2023).

3.3 Greenhouse Gas Emissions from Rice Cultivation

Rice farming significantly contributes to atmospheric CH_4 and N_2O, which are potent greenhouse gases that exacerbate climate change (Bhatia et al. 2010). Nitrogen fertilizer enhances rice growth and promotes the release of organic carbon inputs—such as root exudates, litter, and stubble—which stimulate methanogen activity, further increasing CH_4 emissions (Malyan et al. 2019). The type and quantity of nitrogenous fertilizer applied impacts the amount of CH_4 produced (Rani et al. 2021). Concurrently, ammonia-oxidizing and denitrifying bacteria in paddy soils mediate the nitrogen cycle and are responsible for substantial N_2O emissions (Braker and Conrad 2011). Rising temperatures, driven by climate change, further amplify ammonia volatilization. Alarmingly, approximately 50% of applied nitrogen fertilizer is lost as ammonia (NH_3), nitrous oxide (N_2O), or nitrate (NO_3^-), due to volatilization and microbe-driven nitrification and denitrification processes (Coskun et al. 2017).

Persistent N_2O emissions from rice agricultural systems are a major concern, particularly in South Asia's intermittently flooded fields. Efforts to minimize CH_4 emissions often emphasize intermittent flooding, assuming it yields climate benefits. However, this practice has been shown to increase N_2O emissions relative to conventional flooding (Cowan et al. 2021). As the mitigation of both CH_4 and N_2O is essential in order to meet comprehensive climate targets, "climate-smart" rice cultivation hinges on precision nitrogen management. Rapid nitrification—i.e. the conversion of ammonium (NH_4^+) to nitrate (NO_3^-)—results in inefficient nitrogen use (as plants struggle to assimilate nitrate quickly) and leads to nutrient wastage and lower Nitrogen Use Efficiency (NUE).

The overarching trend highlights N fertilizer's crucial role in feeding a growing population, but also underscores the need for improved NUE to mitigate environmental concerns such as nutrient runoff and greenhouse gas emissions.

3.4 Methane: A Non-CO_2 Greenhouse Gas

Methane (CH_4) is one of the most potent greenhouse gases. With atmospheric concentrations currently around 1.9 ppm, methane's contribution to global warming is significant. Agriculture remains the largest anthropogenic source of this gas, contributing approximately 40% of global human-induced methane emissions (IPCC 2021). Key agricultural contributors include enteric fermentation in livestock, rice cultivation, manure management, and crop residue burning. Over a 100-year timescale, methane is nearly 28 times more effective at trapping heat than carbon dioxide (CO_2), and its impact surges to about 80 times more potent over a 20-year horizon. This elevated Global Warming Potential (GWP) indicates that methane reduction is a critical target in the drive for immediate climate mitigation efforts (Masson-Delmotte et al. 2021).

Methane's climatic impact is amplified by its higher heat-trapping capacity, relatively short atmospheric lifespan, and its accelerated rise in concentration over the past decade. From 2010 to 2019, global CH_4 emissions were estimated at 575 Tg CH_4 per year (Saunois et al. 2025). While natural sources like wetlands play a role, emissions from human activities—especially agriculture, fossil fuel production, and waste management—are growing faster than natural sinks can offset. These sources include-Agriculture accounts for roughly 40% of these anthropogenic methane emissions, fossil fuel operations contribute around 34% and waste management adds approximately 19% (Mundra and Lockley 2024). Given methane's outsized influence on near-term warming, attempts to curb its release into the atmosphere —particularly from agricultural sources must remain—essential aspects of global climate action.

3.4.1 Sources of Methane

Methane Emissions from Livestock

Enteric fermentation is the largest contributor to agricultural methane emissions, accounting for 55–60% of the sector's total. Ruminant animals—such as cattle, sheep, and goats—possess a rumen, a specialized digestive chamber where microorganisms break down food and generate methane, primarily released via burping. In the Indian subcontinent, cattle produce approximately 27 to 58 kg of CH_4 per head annually, while cattle in the U.S. and Europe may emit between 50 and 120 kg CH_4 per head annually (IPCC 2006).

Methane emissions from manure management also occur during the decomposition of manure under anaerobic environments. This typically happens when manure is stored as a liquid or slurry in lagoons or tanks, especially in dairy, beef, and pork production systems. Manure management is responsible for roughly 6–8% of agriculture-related methane emissions. Even manure heaps can release small amounts of methane from their lower, oxygen-deprived layers.

Methane from Paddy Cultivation and Residue Burning

Rice cultivation is the second major source of agricultural methane emissions globally. Flooded rice paddies create anaerobic soil conditions that support methanogenic bacteria, resulting in methane emissions. The gas escapes either through rice plant aerenchyma tissues or as ebullition from the soil, especially during early growth stages. Rice farming contributes about 12–15% of global agricultural methane emissions (Dielmann et al. 2022). Multiple factors influence methane emissions in rice paddies, including water management practices, soil characteristics, rice cultivar traits, and overall farming methods. Key elements that intensify emissions include continuous flooding, high soil organic matter, elevated temperatures, fertilizer application, and the choices of rice varieties.

Crop residue burning, particularly of rice and cereal crops, is a smaller yet notable source of methane emissions in agriculture. It accounts for approximately 3–4% of the total sectoral methane output.

3.5 Mitigation of Methane Emissions

Reduction of agricultural methane emissions represents an essential step toward achieving climate targets. Mitigation requires strategies tailored to specific sources—livestock, manure, and rice cultivation. Beyond direct source mitigation, broader actions like reducing food waste, enhancing agricultural productivity, and enacting supportive policies and incentives are key to achieving sustainable methane reduction in the agricultural sector.

3.5.1 Livestock Methane Mitigation

For livestock, solutions include feed additives (e.g., seaweed, tannins) that inhibit methane-producing microbes, improved nutrition via digestible rations, selective breeding for low-emission animals, and optimized grazing practices (Ku-Vera et al. 2020).

Anaerobic digestion is an effective method to capture methane during manure storage and treatment. It converts methane into biogas, offering both environmental benefits and renewable energy. Complementary techniques for reducing manure-based methane include solid–liquid separation, aerated liquid storage systems, aerobic composting, and covering lagoons to trap methane. These approaches prevent methane escape and improve overall waste handling.

3.5.2 Rice Methane Mitigation

In order to lower methane emissions from rice paddies, the focus has recently been on managing water use, on optimizing organic matter inputs, and on selecting rice varieties with lower methane potential (Malyan et al. 2016).

Techniques for mitigating methane emissions from rice cultivation primarily involve disrupting the anaerobic environments in flooded paddies that promote methane-producing bacteria. Water management strategies are crucial, with Alternate Wetting and Drying (AWD), or intermittent irrigation, proving highly effective. This method involves periodically draining fields to introduce oxygen, which inhibits methane production. Other water-saving approaches like mid-season drainage and Direct Seeded Rice (DSR) also notably reduce emissions. Soil amendments and organic matter management are vital; these include composting crop residues instead of burning or submerging them, using biochar to enhance soil aeration, and employing balanced Integrated Nutrient Management to optimize organic inputs. Furthermore, agronomic practices such as the System of Rice Intensification (SRI) which integrates precise water and nutrient management with other advanced cultivation methods, along with ongoing research into developing low-methane rice varieties, offer additional potential for emission reduction.

While effective at reducing methane, many of these techniques, like AWD and SRI, can have an important implication for N_2O emissions. The shift between aerobic and anaerobic conditions promotes both nitrification (ammonium to nitrate) and denitrification (nitrate to N_2O gas), often leading to increased N_2O fluxes. This creates a trade-off, as a reduction in one potent greenhouse gas (CH_4) might be partially offset by an increase in another (N_2O), with a significantly higher global warming potential (Bhatia et al. 2013).

3.6 Nitrous Oxide (N_2O)

Nitrous oxide (N_2O) concentrations have surged by approximately 123% since pre-industrial times, with the present levels of 331.1 ppb generating increased global concerns. The Kyoto Protocol recognizes N_2O as the third most significant long-lived greenhouse gas. With a GWP of some 265 times greater than CO_2 over a 100-year scale (IPCC 2021), a global temperature potential of 290 compared to CO_2 (Fagodia et al. 2020), and a long atmospheric lifetime (~120 years), N_2O accounts for about 7% of global warming from anthropogenic sources (Lehner 2011). It also contributes to stratospheric ozone depletion, with a radiative forcing of +0.17 Wm^{-2}, making it a major agent after chlorofluorocarbons (CFCs) (Smith 2017). The agriculture, forestry, and land use (AFOLU) sector contribute approximately 24% (equivalent to10–12 Gt CO_2 $eqyear^{-1}$) of global anthropogenic GHG emissions (IPCC 2021), with nitrogen fertilizer and manure being significant sources of N_2O emissions from soils (Hassan et al. 2022).

3.6.1 Sources of Nitrous Oxide

Nitrous oxide (N_2O) arises from both natural and anthropogenic sources, with agriculture being the dominant contributor. Activities like fertilizer application, urine deposition, and manure management are primary sources (Bhatia et al. 2013). As global food demand continues to rise, fertilizer application is projected to triple by 2050 to sustain crop productivity. Besides agriculture, N_2O is emitted from various sectors such as energy, transportation, industrial processes, waste, aviation etc. The sector wise contributions to N_2O emissions, as given in IPCC AR6 report (2021), are shown in Fig. 3.1.

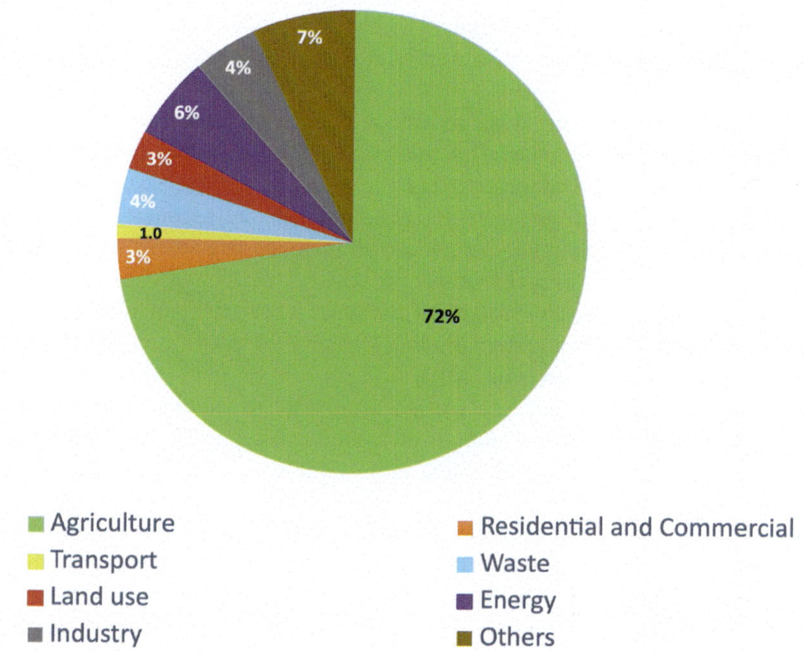

Fig. 3.1 Various source of nitrous oxide (N_2O) [Modified from IPCC 2021]

3.6.2 Nitrous Oxide Emission from Cropped Soil

A significant amount of nitrogen enters agricultural soils globally each year. One study estimated a total nitrogen input of approximately 137 to 145 teragrams (Tg) of N per year (1 teragram = 1 million metric tons). Agricultural practices such as fertilizer use are the main reason for increased N_2O emissions because they create ideal conditions for the soil microbes that produce N_2O. Rice (Oryza sativa), a staple food for nearly half the world's population, supports ~145 million households and consumes roughly 14–15% of global fertilizer use (Heffer et al. 2013). Declining fertilizer efficiency has led to increased fertilizer applications, especially in intensive systems such as the rice–wheat cropping system in the Indo-Gangetic Plain, where nutrient imbalances persist. Excessive nitrogen inputs result in reactive nitrogen losses that intensify N_2O emissions, with N-fertilizer enhancing the availability of organic substrate for methanogenesis (Bhatia et al. 2023). In paddy soils, ammonia-oxidizing and denitrifying bacteria drive N_2O emissions, which escalate with temperature increases. Globally, nearly 50% of applied nitrogen is lost as ammonia (NH_3), nitrate (NO_3^-), or N_2O due to microbial processes (Coskun et al. 2017).

The main factors contributing to N_2O emissions from cropped soils are:

- **Nitrogen Fertilizers**: The widespread application of synthetic nitrogen fertilizers, such as urea and ammonium nitrate, leads to a major source of emissions. When plants can't absorb all the nitrogen, soil microbes convert the leftover fertilizer into N_2O through the processes of nitrification and denitrification (Malyan et al. 2021).

- **Manure and Waste**: Animal waste acts as a nitrogen source when it's used as fertilizer or left to decompose. This added nitrogen, fuels the microbial activity that generates N_2O (Cowan et al. 2021).
- **Crop Debris**: When nitrogen-rich crop residues, especially from legumes, are left in the field to break down, they provide both nitrogen and carbon. This can lead to increased N_2O emissions (Bhattacharyya et al. 2018).
- **Soil Management**: Practices like tilling and other forms of soil disturbance change the soil's air and organic matter content. This creates conditions that favor N_2O production (Bhattacharyya et al. 2018).
- **Water Management**: High levels of soil moisture, particularly in saturated or waterlogged soils, lower the amount of available oxygen. This promotes denitrification, the anaerobic process that is a major source of N_2O (Cowan et al. 2021).
- **Draining Organic Soils**: Draining and farming organic soils like peatlands causes the organic matter to decompose. This process releases large amounts of nitrogen, which in turn leads to high N_2O emissions (IPCC 2006).
- **Paddy cultivation**: Rice fields providing alternating aerobic and anaerobic cycles are significant sources of N_2O emissions, and contribute about 30% to the global N_2O emissions (Dieleman et al. 2022; Mboyerwa et al. 2022).

3.6.3 Mechanism of Soil N_2O Formation

The formation of N_2O in soil involves processes such as nitrification, denitrification, and nitrifier denitrification. These microbial pathways are integral to nitrogen cycling and result in N_2O production under varying soil conditions. As shown in Fig. 3.2. N_2O is generated during microbial oxidation of ammonium (NH_4^+) to nitrate (NO_3^-) in the nitrification pathway (Pathak et al. 2002). Slowing nitrification improves NUE and reduces nitrogen leakage. Nitrification inhibitors (NIs) inhibit enzymes that convert NH_4^+ to NO_3^-, enhancing nitrogen uptake and curbing pollution (Paul et al. 2024). Although moderately priced and beneficial to yields, chemical NIs may accumulate in plants and impact non-target organisms (Ayiti and Babalola 2022). Urease inhibitors (UIs) slow down urea hydrolysis, reducing NH_3 losses and improving fertilization efficiency.

Urea-based fertilizers undergo hydrolysis via urease enzymes, releasing NH_3, which contributes to N_2O emissions through subsequent nitrification. Due to rapid losses via volatilization, denitrification, and leaching, urea exhibits low NUE (Swify et al. 2022). UIs reduce NH_3 losses by inhibiting urease activity (Martins et al. 2017). The combined use of NIs and UIs in enhanced efficiency fertilizers (EEFs) addresses both nitrification and urease pathways, reducing GHG emissions and boosting nitrogen retention. EEFs act as slow-release formulations that improve crop

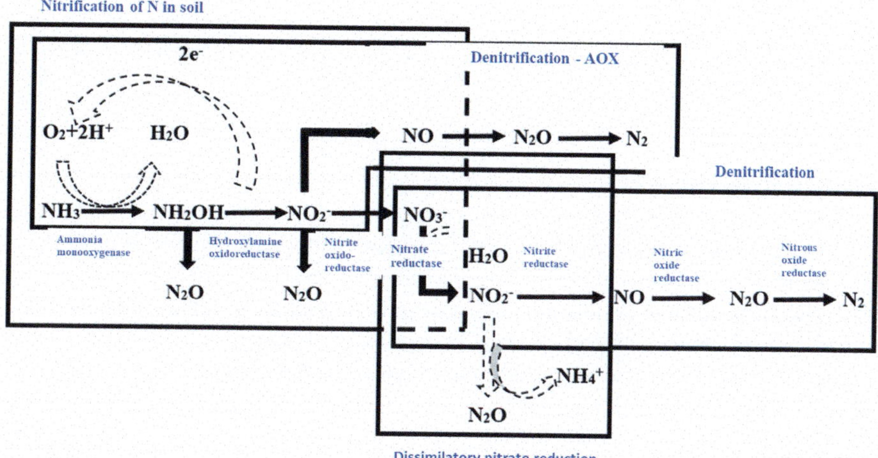

Fig. 3.2 Mineral nitrogen transformation in soil

uptake although they require further field trials to validate their impact. Numerous studies have been performed to evaluate the effectiveness of modified urea fertilizers in reducing NH_3 and N_2O emissions while enhancing NUE in paddy rice cultivation (Paul et al. 2024, 2025).

Nitrification Pathway

In aerobic environments, ammonium (NH_4^+) or ammonia (NH_3) is oxidized to nitrate (NO_3^-), producing intermediates like hydroxylamine (NH_2OH) and nitrite (NO_2^-), with N_2O as a by-product. The conversion involves two groups of bacteria: ammonia oxidizers (e.g., Nitrosomonas europaea) and nitrite oxidizers (e.g., Nitrobacter winogradskyi), collectively known as Nitrobacteriaceae (Wrage-Mönnig et al. 2018). Nitrification supports carbon fixation through energy released during these reactions.

Enzymes facilitate each step: Ammonia monooxygenase converts NH_3 to NH_2OH, Hydroxylamine oxidoreductase transforms NH_2OH to NO_2^-, while Nitrite oxidoreductase catalyzes the conversion of NO_2^- to NO_3^-.

N_2O is also produced via chemical decomposition of NH_2OH and NO_2^- in a process termed chemo-denitrification. Incomplete oxidation of intermediates contributes to additional N_2O release. Aerobic conditions, organic matter availability, and acidic soil pH can amplify N_2O production compared to autotrophic nitrification (Khalil et al. 2002). Environmental conditions heavily influence nitrification rates. For every 10 °C temperature increase, nitrification rates may triple (Breuer et al. 2002). Studies conducted in Southern Australia showed that NIs like N-Serve and DMPP remained effective for up to 42 days under 40–60% water-filled pore space (WFPS) and temperatures between 5 and 15 °C. In contrast, in untreated soils, rising temperature and moisture levels led to significant spikes in N_2O emissions. Application of nitrification inhibitors can lead to significant cumulative emission reductions (Bhatia et al. 2010b).

Denitrification Pathway

Denitrification is a key anaerobic microbial pathway in the nitrogen cycle, involving the sequential reduction of nitrate (NO_3^-) to nitrogen gas (N_2), with nitric oxide (NO) and nitrous oxide (N_2O) as intermediate products (Pathak et al. 2002). The process is mediated by a diverse group of bacteria known as denitrifiers, including genera such as *Thiobacillus*, *Pseudomonas*, *Propionibacterium*, and *Bacillus* (Chen et al. 2013). These facultative anaerobes utilize NO_3^- as an electron acceptor under low oxygen conditions (Wrage-Mönnig et al. 2018). Enzymatic catalysis plays a central role: nitrate reductase converts NO_3^- to nitrite (NO_2^-); nitrite reductase reduces NO_2^- to NO; nitric oxide reductase transforms NO to N_2O; and nitrous oxide reductase finally converts N_2O to N_2.

Denitrification has drawn global attention due to its role in contributing to atmospheric N_2O, a potent greenhouse gas. Although the process helps remove excess nitrate and nitrite from soils, incomplete denitrification results in the release of N_2O instead of N_2 (Albertsson et al. 2019). Several environmental factors influence this emission. For instance, low soil pH inhibits nitrous oxide reductase activity, limiting N_2 formation and increasing N_2O output. Higher concentrations of NO_3^- and O_2 further elevate the N_2O/N_2 ratio. While NO_3^- competes with N_2O as an electron acceptor, elevated O_2 levels disrupt the anaerobic conditions needed for denitrification, suppressing enzyme activity and nitrate reduction (Chen et al. 2013).

In waterlogged environments, such as those caused by flooding or thawing permafrost, denitrification plays a critical role by releasing N_2O.

Nitrifier Denitrification

Nitrifier denitrification is favoured in environments like saturated agricultural soils or wastewater treatment systems. It is a metabolic pathway used by certain nitrifying microbes (specifically, ammonia-oxidizing bacteria, AOB) where they shift from their normal aerobic process to an anaerobic-like process to survive under low-oxygen conditions and low organic carbon. Under limited oxygen the AOB oxidizes NH_3 to NO_2^- as usual. However, because oxygen is scarce, the AOB switches to using the NO_2^- it just produced. The AOB reduces the internal nitrite through a series of steps (similar to the canonical denitrification pathway) in production of N_2O gas from NO_3^-.

In simple terms, it's a "crossover" pathway where an organism known for one job (nitrification) temporarily performs a step of another job (denitrification) using its own oxidized nitrogen compounds.

3.7 Drivers of Rising Nitrous Oxide Emissions

Recent increases in atmospheric N_2O have been largely attributed to the expansion in agricultural lands and to intensified use of synthetic nitrogen fertilizers (Reay et al. 2012). Activities such as excessive fertilizer application and unmanaged manure use contribute to elevated emissions. Overuse of nitrogenous fertilizers, coupled with poor uptake efficiency by crops, leads to increased N_2O emissions, denitrification, and nutrient loss through leaching and runoff (Kumar et al. 2025). These processes reduce nutrient efficiency, generate economic losses, and contribute to environmental degradation (Gupta et al. 2016). Fertilizers and organic inputs promote microbial nitrification and denitrification of mineral nitrogen, releasing N_2O into the atmosphere. Organic amendments serve as sources of organic carbon stimulating the microbial growth and nitrogen assimilation. This microbial activity may temporarily reduce N_2O production by increasing competition for ammonium (NH_4^+) between autotrophic and heterotrophic organisms (Chen et al. 2013).

3.8 Mitigation of Nitrous Oxide Emissions

Various mitigation strategies have been proposed and implemented. These include the use of nitrification inhibitors, slow-release or controlled fertilizers, conservation tillage, and optimized water and fertilizer management (Bhatia et al. 2023).

Strategies for mitigating N_2O emissions from cropped soils are primarily centered on optimizing nitrogen use and managing soil conditions. A key approach is fertilizer management, which focuses on applying nitrogen more efficiently. This involves use of the "4R nutrient stewardship framework" applying the right amount of fertilizer at the right time, in the right place, and using the right source (Bhatia et al. 2012b).

The leaf colour chart (LCC) is an easy-to-use and affordable diagnostic device for observing the relative greenness of a rice leaf as an indicator of plant N status. The latter is closely linked to photosynthetic rate and biomass production and is a sensitive indicator of differences in crop N demand during a growing season. A tool to rapidly estimate leaf nitrogen status and thereby guide the application of fertilizer to maintain optimal leaf greenness is crucial for attaining a high rice yield with effective N management (Bhatia et al. 2012b). The initial prototype of LCC was developed by IRRI and the Philippines Rice Research Institute (PhilRice). Subsequently several other institutes developed the LCC technique based on their region and on other parameters. In Practical terms, an LCC chart has four green strips (Fig. 3.3), with color ranging from yellow green to dark green. A simple comparison determines the greenness of the rice leaf, which indicates its N content.

There are several fertilizer calculators developed by different organizations according to region and the specific crop and based on the irrigation situation of that crop. Among these, the ICAR-Central Coastal Agricultural Research Institute, based in Goa, India developed their Fertilizer Calculator. This is a completely offline

Fig. 3.3 Comparison of leaf nitrogen content with the LCC colour strips

app (Fig. 3.4) for Android phones (termed the "Soil Test-Based Fertilizer Recommender", STFR) which can perform calculations according to the area of the farm or the number of plants/trees (Mahajan 2021). If users adhere to the calculator's suggestions, they can prevent or reduce fertilizer input loss and ensure that crops are appropriately nourished. This app is flexible and can be easily adapted to other regions, crop variety, and soil fertility levels.

In addition to fertilizer management, water and soil management are crucial for reducing N_2O emissions. Since, N_2O production is favored in waterlogged, low-oxygen environments, practices like drip irrigation and improved drainage can contribute by restricting the development of these conditions. Likewise, reducing soil disturbance through no-till farming and using cover crops both help to maintain a healthy soil structure and absorb excess nitrogen, thereby minimizing the amount available for conversion into N_2O. Enhanced-efficiency fertilizers, which include nitrification and urease inhibitors and controlled-release products, are also highly

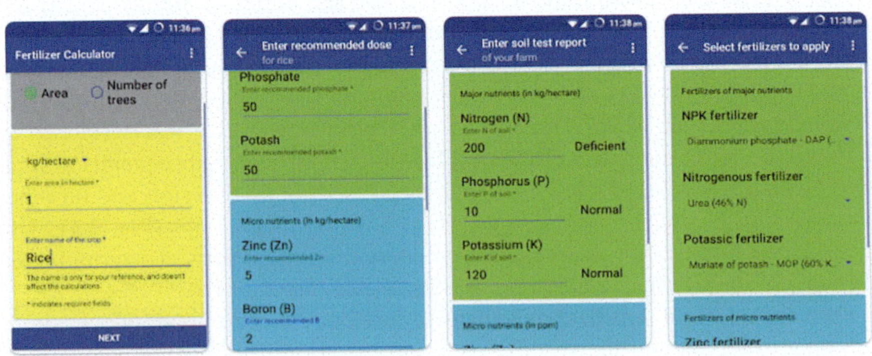

Fig. 3.4 Practical application of the STFR fertilizer calculator for a rice crop

effective as they slow down the conversion of nitrogen into forms that are easily lost to the atmosphere. A combination of these various approaches can significantly reduce emissions and improve overall agricultural sustainability.

The next chapter focuses specifically on the role of enhanced-efficiency fertilizers in reducing N_2O and ammonia (NH_3) emissions from agricultural soils.

Chapter 4
Efficient Fertilizers

4.1 Enhanced Efficiency Fertilizers

Enhanced Efficiency Fertilizers (EEFs) refer to specialized fertilizer formulations designed to improve nutrient uptake by plants while minimizing environmental losses through runoff, gaseous emissions, and leaching (Association of American Plant Food Control Officials AAPFCO 2013). Compared to conventional fertilizers, EEFs offer notable environmental, agronomic, and economic advantages (Paul et al. 2024).

The concept of EEFs dates back to 1924, and was marked by the introduction of urea–formaldehyde an early example of a slow-release nitrogen product (Timilsena et al. 2015). Research indicates that EEFs maintain consistent nutrient availability throughout the growing season, thereby promoting sustained crop growth. These formulations reduce stress-related damage such as leaf burn and osmotic stress, thereby improving plant resilience and seasonal growth patterns (Shaviv 2001; Trenkel 2021). In addition, their hygroscopic nature makes EEFs more suitable for long-term storage, unlike conventional nitrogen fertilizers that tend to cake due to moisture absorption.

EEFs also help to reduce emissions of nitrous oxide (N_2O) and ammonia (NH_3), contributing to both environmental protection and economic benefits for farmers by lowering fertilizer usage and minimizing losses (Timilsena et al. 2015). Common examples include controlled release Controlled Release Fertilizers (CRFs) with urease inhibitors (UIs), and nitrification inhibitors (NIs), which have been developed to enhance nutrient use efficiency (NUE) (Akiyama et al. 2010). Their point of intervention in the N mineralization pathways can be understood from Fig. 4.1.

U. C. Kulshrestha et al., *Atmosphere-Biosphere Interactions of Reactive Nitrogen in South Asia*, Springer Briefs in Interdisciplinary Geosciences – Asia-Pacific,
https://doi.org/10.1007/978-3-032-17194-8_4

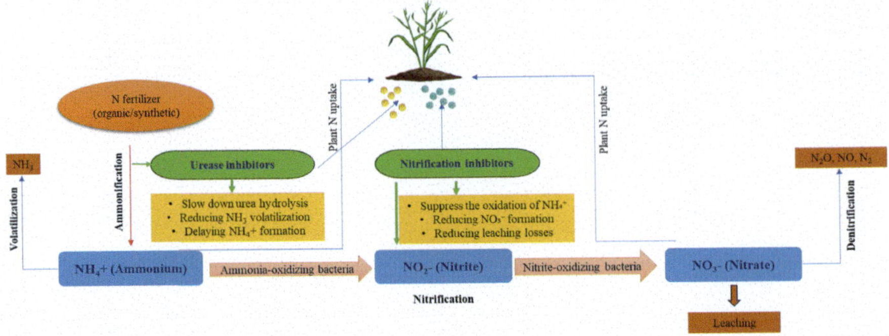

Fig. 4.1 The role of nitrification and urease inhibitors role in nitrogen mineralization pathways

4.2 Effectiveness of Enhanced Efficiency Fertilizers in Agricultural Systems

Nitrogen loss is a major concern in agricultural fields, often leading to environmental degradation and reduced crop productivity. To mitigate these issues, effective management of nitrogen fertilizers is essential. Studies have shown that EEFs can lead to an 8–10% increase in nitrogen uptake and a 5–10% improvement in rice yield.

Soil pH plays a critical role in determining the effectiveness of EEFs. In alkaline soils (pH $\geq$ 8), yield response improves by up to 10%, while in acidic soils show negligible impact on nitrogen uptake or crop yield (Linquist et al. 2013).

4.3 Environmental and Ecosystem Benefits of EEFs

EEFs offer several environmental and ecosystem benefits by optimizing nutrient delivery and reducing losses to the environment (Fig. 4.2). These benefits include:

- **Reduced Nutrient Runoff and Leaching**: EEFs, particularly controlled-release and slow-release types, release nutrients gradually. This minimizes the amount of soluble nutrients in the soil at any given time, thereby reducing the risk of excess nutrients being washed away by rain (runoff) or seeping into groundwater (leaching). This helps protect surface water bodies from eutrophication and prevents groundwater contamination.
- **Lower Greenhouse Gas Emissions**: Some EEFs can help reduce emissions of potent greenhouse gases, such as nitrous oxide (N_2O). For instance, nitrification inhibitors can slow down the conversion of ammonium to nitrate, a process that can lead to N_2O emissions. By ensuring that more nitrogen remains in the form of ammonium, (which is less prone to denitrification), EEFs can mitigate these emissions.

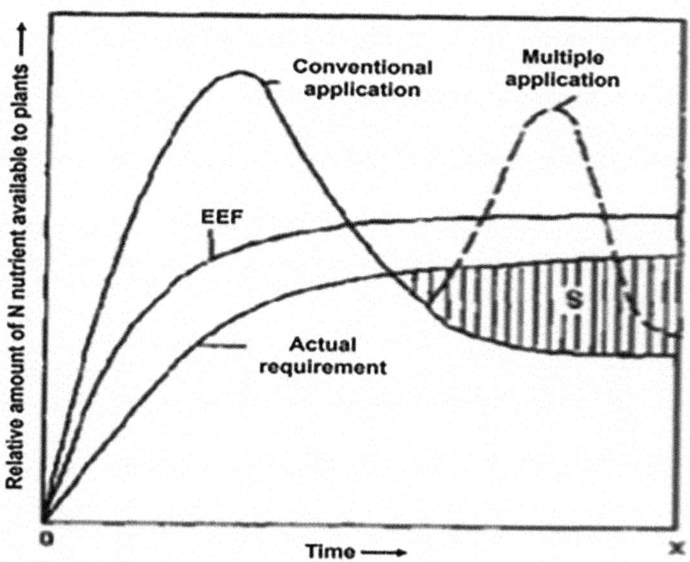

Fig. 4.2 Modified availability of N nutrient to plants when using conventional fertilizer and Enhanced Efficiency Fertilizer (EEF)

- **Improved Air Quality**: By reducing ammonia (NH_3) volatilization, especially from urea-based fertilizers, EEFs contribute to better air quality. Since ammonia emissions can contribute to particulate matter formation and acid rain. Urease inhibitors, a type of EEF, slow down the enzymatic breakdown of urea to ammonia, thus reducing these losses to the atmosphere.
- **Enhanced plant N uptake and Nutrient Use Efficiency (NUE):** EEFs improve the synchronization between nutrient availability and plant demand. Figure 3.4 provides a schematic diagram of nutrient demand from plant during its growth cycle which illustrates the differences between conventional fertilizer application and the actual requirement of the plant. In the case of EEF use, more of the applied nutrients are taken up by the crops, leading to less waste and a reduced need for multiple fertilizer application. Higher NUE translates into fewer nutrients escaping into the environment.
- **Reduced Eutrophication**: By minimizing nitrogen and phosphorus runoff into water bodies, EEFs help combat eutrophication. Eutrophication (the over-enrichment of water with nutrients) leads to excessive algal growth, oxygen depletion, and harm to aquatic life.
- **Preservation of Biodiversity**: By reducing nutrient pollution in aquatic and terrestrial ecosystems, EEFs help maintain the health and biodiversity of these environments. Reduced nutrient imbalances prevent the dominance of certain species that thrive onexcess nutrients, thereby preserving the natural balance of ecosystems.

- **Sustainable Resource Management**: By making fertilizer use more efficient, EEFs support more sustainable agricultural practices (Reay 2012). They help conserve finite mineral resources used in fertilizer production and reduce the energy consumption associated with fertilizer manufacturing and transportation due to potentially lower application rates.

4.4 Enhanced Efficiency Fertilizers Improves Crop Yield and Reduces N_2O Emissions

Substantial evidence supports the role of Enhanced Efficiency Fertilizers (EEFs) in improving crop productivity while simultaneously reducing nitrous oxide (N_2O) emissions. However, their effectiveness is influenced by factors such as soil type and agricultural management practices. A meta-analysis by Thapa et al. (2016) evaluated the impact of EEFs—including controlled-release fertilizers (CRFs), nitrification inhibitors (NIs), and dual inhibitors (DIs combining nitrification and urease inhibition) on three major cereals: rice, wheat, and corn. The study found consistent reductions in N_2O emissions of 19%, 38%, and 30% respectively for CRFs, NIs, and DIs compared to conventional nitrogen fertilizers.

NIs were particularly effective in acidic soils, whereas DIs performed better in alkaline and coarse-textured soils. Urease inhibitors (UIs) showed greater efficacy in coarse soils when coupled with effective irrigation practices. Importantly, both NIs and DIs demonstrated a dual benefit: reducing both N_2O emissions and improving crop yields. These findings highlight the fact that optimizing soil conditions and management strategies enhances the performance of EEFs.

Further support for this approach comes from a similar meta-analysis by Akiyama et al. (2010), which showed that polymer-coated fertilizers (PCFs) and NIs effectively reduced emissions of both N_2O and nitric oxide (NO) from organic and synthetic sources. While NIs maintained consistent performance across various land uses and soil types, PCFs exhibited selectivity depending on conditions. Due to their reliability and broad applicability, NIs stand out as a dependable option for mitigating nitrogen-related greenhouse gas emissions.

4.5 Enhanced Efficiency Fertilizers Reduce Ammonia (NH_3) Emissions

Beyond reducing N_2O emissions, EEFs are also instrumental in curbing ammonia (NH_3) volatilization. In a comparative study, Connell et al. (2011) assessed the performance of various EEF formulations including maleic-itaconic copolymer-treated urea (MICPU), N-(n-butyl) thiophosphorictriamide (NBPT), dicyanamide-treated urea, NBPT-treated urea, and polymer-coated urea (PGU) against conventional nitrogen sources like urea, ammonium nitrate (AN), and enhanced efficiency urea-ammonium nitrate (UAN) blends. Research conducted on bermudagrass (*Cynodon-dactylon*), revealed a marked reduction in NH_3 emissions when urea was treated with NBPT. This formulation also achieved comparable agronomic performance to UAN and AN. Additionally, PGU was found to improve nitrogen concentration in forage while decreasing NH_3 volatilization. The NBPT application not only boosts urea's nutrient efficiency but also mitigated NH_3 emissions making it a valuable tool for sustainable nitrogen management.

4.5.1 Enhanced Efficiency Fertilizers Reduce Mineral N Leaching Losses

Enhanced Efficiency Fertilizers (EEFs) significantly reduce mineral nitrogen (N) leaching by controlling nutrient release and optimizing its availability to plants. Through mechanisms such as polymer coatings, which allow for slow diffusion of nutrients, or inhibitors that temporarily block the conversion of ammonium to more mobile nitrate, EEFs minimize the amount of soluble nitrogen vulnerable to being washed through the soil profile and into groundwater or surface waters. Studies have shown that EEFs can substantially decrease nitrate–N leaching, with some reports indicating reductions of 17–58% or even up to 75% under specific conditions. Ultimately these active properties improve nutrient use efficiency and mitigate.

4.5.2 Agronomic Benefit of Enhanced Efficiency Fertilizers

The effectiveness of these fertilizers in rice cultivation can be assessed across several agronomic dimensions:

- **Nitrogen Uptake**: Slow-release fertilizers provide nitrogen in sync with crop demand, boosting uptake. Measuring N concentration in leaves, stems, and grains helps gauge this efficiency.
- **Yield**: Consistent nutrient availability improves crop performance. When aligned with the specific nutrient requirements and growing conditions, slow-release fertilizers often lead to higher yields.
- **Morphological Characteristics**: Steady nutrient release positively influences plant traits such as height, number of tillers, and leaf area, which are key indicators of healthy development.
- **Harvest Nitrogen Use Efficiency (NUE)**: This metric evaluates nitrogen use for total biomass production. Slow-release fertilizers can substantially improve biomass yield and boost the utilization of applied nitrogen.

- **Grain Nitrogen Use Efficiency**: This indicator reflects nitrogen utilization for grain formation. By maintaining nutrient availability during critical reproductive stages, slow-release fertilizers can significantly enhance grain NUE.

4.6 Slow-Release and Controlled Release Fertilizers

Slow-release urea formulations such as sulfur-coated urea, lac-coated urea, polymer-coated urea, and urea "super granules" have been widely studied for their ability to reduce nitrogen losses. Additionally, various hydrolysis-retarding chemicals have also been explored (Gould et al. 1986). Although the terms *slow-release fertilizer* (SRF) and *controlled-release fertilizer* (CRF) are often used interchangeably, their operational mechanisms differ. SRFs are typically low-solubility compounds with complex or high molecular weight structures that release nutrients via microbial or chemical decomposition (Shaviv 2001). In contrast, CRFs consist of water-soluble nutrients encased in a coating that regulates their release into the soil.

SRFs are broadly categorized into urea-aldehyde condensation products, physically "barriered" fertilizers (i.e. fertilizers employing a physical barrier either coated or matrix-incorporated), and super granules. CRFs fall under the subset of SRFs with physical coatings. Although nutrient release timings for CRFs depends on individual plant species and their metabolic demands, the European Standardization Committee (CEN) Task Force has proposed guidelines for CRF performance. According to these standards:

- No more than 15% of nutrients should be released within the first 24 h.
- No more than 75% should be released by day 28.
- A minimum of 75% should be released within the designated release timeframe

These benchmarks ensure CRFs deliver nutrients gradually and predictably. Additional expectations from CRFs include affordability, environmental compatibility, and sustainability (Trenkel 2021).

Slow-release and controlled-release fertilizers are formulated to supply nutrients steadily over time, improving both nutrient use efficiency and crop performance. In sugarcane cultivation, sulfur-coated fertilizers have demonstrated notably lower N_2O emissions (30 g ha^{-1} d^{-1}) than conventional urea (Soares et al. 2015). In rice, they have been linked to enhanced grain yields and improved nitrogen efficiency. Components such as phosphogypsum and bentonite sulfur are often used to enhance fertilizer functionality (Paul et al. 2024).

Table 4.1 Differences between the CRF and SRF techniques

Feature	Slow-release fertilizer (SRF)	Controlled-release fertilizer (CRF)
Nutrient release	Less control over release; influenced by multiple factors	More predictable and controlled release; less influenced by soil conditions
Influencing factors	Soil temperature, moisture, pH, and microbiology	Primarily soil temperature; less dependent on soil microbiology
Mechanism	Reduced solubility of inorganic compounds or slow diffusion	Coating with a material that controls nutrient release over time
Benefits	Delays and limits nutrient availability; can reduce environmental losses	Mimics plant nutrient demand more closely, reducing losses and increasing yield
Application	Fewer applications, often used for established lawns or crops with steady nutrient needs	Single applications can provide nutrients for extended periods, suitable for various crops
Examples	Sulphur coated urea, Chicken manure	Urea–formaldehyde, Osmocote

4.7 Distinction Between Controlled-Release Fertilizers and Slow-Release Fertilizers

While both CRFs and SRFs are applied to soil to facilitate gradual nutrient availability, their modis operandi differ significantly. To repeat despite frequent interchangeable use of the terms, they remain distinct approaches and are not synonymous, as demonstrated in the comparative analysis (presented in Table 4.1).

4.8 Controlled Release Fertilizers: Design and Mechanism

Controlled-release fertilizers (CRFs) are designed to gradually provide nutrients to plants gradually over an extended period, in contrast to conventional fertilizers that release nutrients quickly. This controlled release helps to synchronize nutrient availability with plant needs, improve nutrient use efficiency, and reduce environmental pollution.Controlled release fertilizers are specially engineered granules formulated with excipients that act as carrier molecules (Fig. 4.3). These carriers regulate the gradual release of nutrients, enhancing nutrient use efficiency (NUE) and minimizing environmental losses from leaching, runoff, or volatilization (Malla et al. 2005). The thickness of the coating applied to these formulations directly affects the duration and rate of nutrient release, offering higher efficiency than traditional fertilizers and reducing the need for repeated applications (Timilsena et al. 2015).

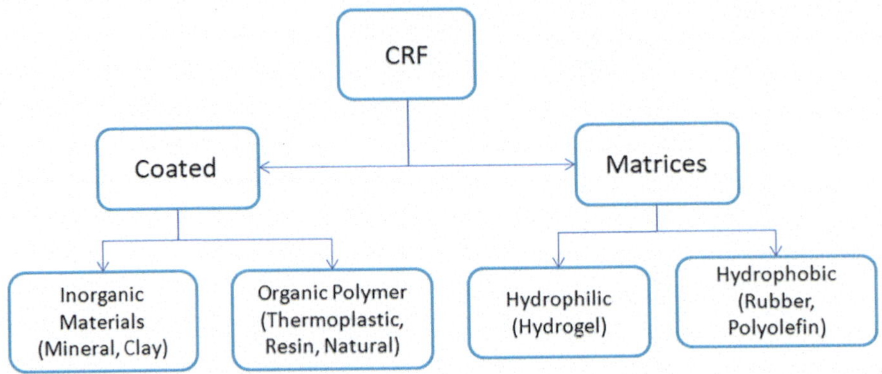

Fig. 4.3 Controlled-release fertilizers

4.9 Classification of Controlled-Release Fertilizers

The CRFs (Fig. 4.3) can be categorized into three major groups (Shaviv 2001; Trenkel 2021)

(a) **Organic Compounds**: These include: (i) synthetic organic nitrogen compounds with low solubility fertilizers that decompose chemically (e.g., cyclo-diurea, isobutylidenediurea) or biologically (e.g., urea formaldehyde) and (ii) natural organic compounds from animal manure and sewage sludge, contributing nutrients through microbial decomposition.

(b) **Water-Soluble Compounds with Physical Barriers including Materials that Restrict Nutrient Dissolution**: These appear either as matrices of active nutrients embedded in hydrophobic materials (e.g., hydrogels, rubber, polyolefins) or as granules coated with polymers classified as inorganic coatings (mineral or sulfur-based) or as organic coatings (resins, thermoplastics)

(c) **Inorganic Low-Solubility Compounds:** These include mineral-based materials such as acidulated phosphate rock and metal ammonium phosphates (e.g., magnesium ammonium phosphate, $MgNH_4PO_4$).

4.10 Operating Mechanisms of Controlled Release Fertilizers

The primary mechanisms and modes of action of CRFs generally involve a physical barrier or a modified chemical structure that regulates the release rate. CRFs are designed to release nitrogen in alignment with the plant's metabolic demands. The coating acts as a barrier, preventing unregulated nutrient discharge (Shaviv 2001). Waterborne starch biopolymers, especially when modified with polyvinyl alcohol, serve as cost-effective and environmentally friendly coating materials. Studies have shown that thicker coatings prolong nutrient release time while decreasing the diffusion coefficient. For optimal CRF performance, coatings must exhibit uniformity and robust film integrity (Paul et al. 2024).

4.11 Coated Fertilizers

These are granular fertilizer particles encapsulated with a protective layer, often made of polymers, sulfur, or waxes. The release of nutrients occurs primarily through diffusion of dissolved nutrients through the coating and/or degradation of the coating material. The effectiveness of Controlled Release Fertilizers (CRFs) is closely tied to their underlying release mechanism. While several models have been proposed, the multi-stage diffusion model is the most widely accepted explanation (Shaviv 2001). In this model, a CRF is conceptualized as a coated granule with a solid nutrient core. Upon application to the soil and exposure to irrigation, water penetrates the protective outer coating and condenses within the solid core. This initiates the partial dissolution of nutrients inside the granule. As dissolved solutes accumulate, osmotic pressure builds within the fertilizer particle, causing it to swell. Depending on the structural properties of the coating and environmental conditions, this swelling leads to one of two pathways for nutrient release:

Release by Diffusion mechanism: Here the coating acts as a barrier, that prevents immediate dissolution and provides a slow, sustained release of nutrients. The water permeates through the semi-permeable coating, dissolving the fertilizer inside. This release mechanism functions effectively when the membrane's threshold resistance is sufficient to withstand the osmotic pressure buildup within the fertilizer granule. Under these conditions, nutrients are released gradually through a controlled diffusion process, which takes place slowly through the coating into the soil. The driving force behind this diffusion may stem from a pressure gradient, a concentration gradient, or a combination of both. The rate of diffusion is influenced by the thickness and composition of the coating, as well as soil temperature and moisture. Fertilizers coated with polymers—such as polyolefin—typically utilize this diffusion-based release method. The coating material may also degrade over time due to microbial activity, temperature, or moisture, with the result the enclosed nutrients are gradually released.

Release as a result of Membrane Failure: If the membrane's threshold resistance fails to withstand the increasing osmotic pressure within the fertilizer granule, the outer coating may rupture, resulting in the abrupt and uncontrolled release of the entire nutrient core. This phenomenon is known as *catastrophic release*. Water can enter the coating, causing the fertilizer inside to dissolve and increasing osmotic pressure, which can then force nutrients out through pores or small imperfections in the coating. It is commonly observed in fertilizers coated with sulfur or modified sulfur, which tends to produce relatively fragile and less resilient coatings (Fig. 4.3).

4.12 Matrix-Based Fertilizers

These involve incorporating fertilizers into a matrix material, such as a polymer or gel. The nutrients are embedded within the matrix, and their release is controlled as the matrix material degrades or swells, allowing nutrients to slowly become available. The matrix physically holds the nutrients, releasing them as the matrix breaks down or releases them through its structure (Fig. 4.4).

Nutrient release from controlled release fertilizers (CRFs) is influenced by a range of environmental factors. Research indicates that elevated temperatures combined with high moisture levels significantly accelerate nutrient release rates (Trenkelet al. 2021). This highlights the pivotal role of water as the medium that facilitates the movement of nutrients from the fertilizer-polymer interface to the polymer–soil interface, thereby enabling their availability to crops (Azeem et al. 2014). Some CRFs achieve slow release through their chemical composition, often by having low solubility or requiring microbial or chemical degradation to become available. Examples include urea–formaldehyde or isobutylidenedi urea (IBDU), which slowly convert to plant-available forms in the soil.

4.13 Factors Influencing Release Rate

The factors that affect the rate of nutrient release from CRFs are the coating material, thickness, and permeability. Soil temperature and moisture also effect the release rate. Higher temperatures and moisture generally increase the release rate. Microbial activity within the soil is especially important relevant for coatings that degrade microbially. Soil pH and texture can also influence degradation and diffusion of CRFs into soil. The amounts of irrigation and rainfall affect the water availability for dissolution and diffusion.

4.14 Controlled-Release Fertilizers and Their Role in N_2O Mitigation

The predictable nutrient release profile of CRFs makes them ideal for mitigating nitrous oxide (N_2O) emissions from agricultural soils. Halvorson et al. (2014) evaluated several Enhanced Efficiency Fertilizers (EEFs) in irrigated corn (*Zea mays L*) cultivated on clay loam soil. Notably, Environmentally Smart Nitrogen (ESN), a polymer-coated urea CRF, reduced N_2O emissions by 42% compared to conventional urea and by 14% relative to urea–ammonium nitrate (UAN) under no-till and strip-till management. Furthermore, UAN blended with a slow-release nitrogen fusion demonstrated even greater effectiveness, lowering N_2O emissions by 57% and 28% compared to urea and UAN respectively.

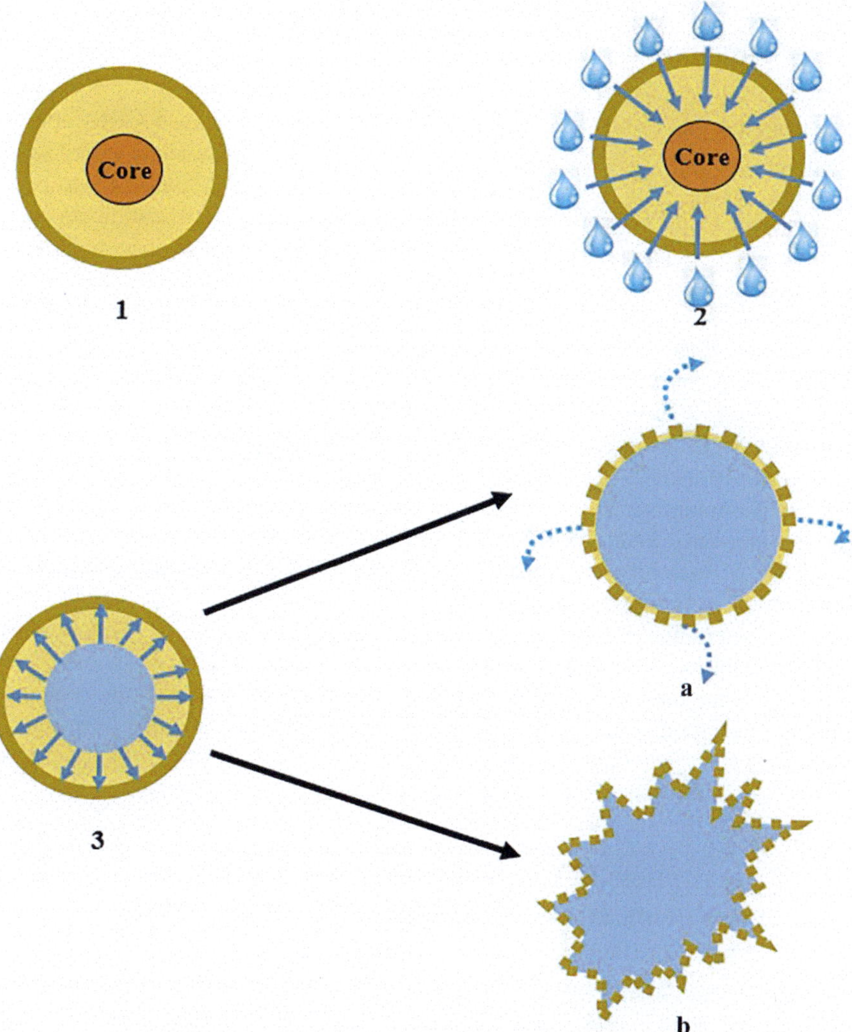

Fig. 4.4 Multi stage diffusion model of CRFs 1. Fertilizer with polymer coating and solid core. 2. Irrigation water penetrates through the outer coating and enters the core, 3. Development of osmotic pressure resulting in fertilizer dissolution: **a** Diffusion-release mechanism, **b** Membrane Failure mechanism

4.15 Soil Moisture and Controlled-Release Fertilizer Performance

Several laboratory studies have examined the interaction between CRFs and soil moisture levels. One such experiment used andosol soils maintained at varied water-filled pore space (WFPS) levels specifically 40%, 55%, 70%, and 85% to represent a range of conditions under which nitrification and denitrification processes contribute to N_2O emissions. Three polymer-coated fertilizers -Nutricote Type 40, containing nitrogen in the forms of ammonium phosphate and ammonium sulfate, were compared against standard soluble fertilizers such as calcium nitrate and ammonium sulfate.

The experimental data revealed that ammonium sulfate resulted in higher N_2O emissions than calcium nitrate, indicating that nitrification was the dominant emission pathway. Emissions peaked at 85% WFPS, but application of CRF with ammonium-nitrogen (NH_4^+–N) delayed and suppressed these peak emissions. Likewise, CRFs containing NO_3^-–N yielded lower emissions than conventional NO_3^- fertilizers but only under WFPS levels below 85%. These results emphasize the importance of matching fertilizer types to prevailing soil moisture conditions in order to effectively reduce emissions.

Similar outcomes have been documented in field applications. Thermoplastic resin-coated urea applied in paddy fields significantly reduced N_2O emissions, particularly during the mid-season aeration phase. This CRF also lowered the concentration of N_2O trapped within soil–water pore spaces, further confirming its mitigation potential under flooded cropping conditions.

4.16 Effectiveness of Controlled-Release Fertilizers in Reducing N_2O Emission from Tea Cultivation

Field experiments conducted on tea (*Camellia sinensis*) have demonstrated the superior performance of sulfur-coated urea a type of CRF in mitigating nitrous oxide (N_2O) emissions. Compared to organic fertilizers composed of chicken manure and oil cake, sulfur-coated urea was notably more effective in reducing greenhouse gas release from the field (Deng et al. 2017).

4.17 Role of Controlled-Release Fertilizers in Reducing Ammonia (NH_3) Volatilization

Ammonia volatilization and nitrogen surface runoff are widespread issues in rice (*Oryza sativa* L.) cultivation, contributing significantly to water and air pollution. Numerous studies have evaluated these nitrogen losses alongside metrics such as nutrient use efficiency (NUE) and crop yield. Li et al. (2018) found that CRFs specifically polyurethane-coated urea and degradable polymer-coated urea reduced nitrogen runoff by 8–58% and NH_3 volatilization by 23–62% compared to conventional urea. These applications also led to decreased surface water pH and lower ammonium-nitrogen (NH_4^+–N) concentrations, contributing to improved environmental outcomes. In addition to reduced emissions, CRF application significantly increased grain yield, NUE, and nitrogen uptake. Some studies reported ammonia volatilization reductions of 51.34% and 91.34% with resin-coated urea and sulfur-coated urea, respectively. These fertilizers also enhanced photosynthetic rates, which likely contributed to improved NUE.

4.18 Nitrification Inhibitors (NIs) as Enhanced Efficiency Fertilizers (EEFs)

Nitrification inhibitors (NIs) have shown promising potential in reducing nitrogen emissions from both organic and mineral fertilizers. Their primary mode of action involves suppressing ammonia monooxygenase (AMO) the enzyme responsible for converting ammonium (NH_4^+) to nitrate (NO_3^-) thereby limiting both nitrification and subsequent denitrification pathways that lead to nitrous oxide (N_2O) emissions (Malla et al. 2005). Well-established NIs such as 3,4-dimethylpyrazole phosphate (DMPP) and dicyandiamide (DCD) have demonstrated N_2O emission reductions of 40.2% and 42.3% respectively. Their corresponding emission factors were measured at 0.7% and 0.41%. Meta-analyses have revealed the inhibitory impacts across various land uses, with the highest effect found on grassland, followed by cropland, upland and paddy, respectively (Gilsanz et al. 2016).

Ammonia oxidation—a crucial step in nitrification—is governed by two distinct microbial groups: ammonia-oxidizing archaea (AOA) and ammonia-oxidizing bacteria (AOB) (Yue et al. 2021). Their differences in lipid structure and AMO gene sequences influence the resulting sensitivity to inhibitors. For example, *Nitrosospiramultiformis* (AOB) and *Ca. Nitrososphaeraviennensis* (AOA) exhibit differing responses to allylthiourea, DCD, and amidinothiourea, with AOA showing lower sensitivity (Shen et al. 2013). Interestingly, half maximal effective concentration (EC_{50}) for allylthiourea was 1000 times higher for AOA. *Nitrosotaleadevanaterra*, an AOA species, was less susceptible to allylthiourea but responded similarly to nitrapyrin and DCD at concentrations comparable to AOB. Understanding these microbial distinctions allows targeted strategies for emission mitigation.

4.19 Urea Deep Placement for Enhanced Efficiency Fertilization

Urea deep placement (UDP) is a highly efficient, innovative and proven nitrogen management technology that has revolutionized rice cultivation in many Asian countries, particularly in Bangladesh. UDP is a fertilizer application technology that achieves average yield increases of 15–20% and urea use decreases by 25–35% (Gregory et al. 2010; Miah et al. 2015; IFDC, 2016). The UDP technique decreases N losses by 40% and increases urea use efficiency to 30–40% compared to broadcasting of prilled urea (Gregory et al. 2010; Miah et al. 2015; Huda et al. 2016). The latter is the usual technique of urea fertilizer application but is a very inefficient practice, with a 60 to 70% loss of the applied N, and giving further contributions to GHG emissions and water pollution. In the UDP technique, urea is made into briquettes of 0.9 to 2.7 g in size which are placed at 7 to 10 cm soil depth near the root zone after rice transplantation. Urea briquettes are applied between 4 hills of rice at alternate rows only once during the crop growth season, compared with two to three fertilizer applications needed for broadcasting of prilled urea (Fig. 4.5). UDP releases N slowly and makes N available to the crop throughout the growing periods. As plants can then efficiently uptake N from the soil, this results in higher nitrogen use efficiency (NUE) and crop yields, and lower N losses and production costs.

The UDP approach is recognized as a highly efficient N management practice in rice cultivation that contributes to sustainable agricultural intensification and supports multiple Sustainable Development Goals (SDGs), including hunger eradication, clean water availability, responsible consumption, and climate action. The technology enhances farmers' incomes by reducing fertilizer use and boosting yields while also protecting the environment through lower nitrogen losses, reduced greenhouse gas emissions, and a more sustainable use of agrochemicals. Despite these clear advantages, adoption by farmers level remains limited due to technical, socio-economic, and institutional challenges (Alam et al. 2023).

Fig. 4.5 Single row urea super granule, and prilled urea applicators developed by Bangladesh Agricultural Research Institute (BARI 2018)

A major barrier to adoption is the labor-intensive nature of USG application, as it requires placement at a depth of 7–10 cm deep between rice hills, unlike the simpler broadcast method. While applicators have been developed, they remain scarce and costly in rural areas. Moreover, the available applicators are inefficient at placing USG in the root zone. Limited farmer awareness about long-term benefits, along with preference for traditional methods, further restricts adoption. Accessibility issues also play a role since prilled urea, being subsidized and widely available, is more attractive to farmers. In addition, labor shortages and high wages during transplanting make USG placement difficult and costly. Socio-economic constraints, including small landholdings, limited capital, and risk aversion, discourage experimentation with new methods that demand greater initial effort. Institutional weaknesses, such as inadequate extension services and insufficient policy support, compound the problem by failing to demonstrate USG's practical benefits effectively (Ajibola and Fatoki 2017). Finally, entrenched cultural habits and resistance to change further slow uptake. Collectively, these barriers explain the limited adoption of USG despite its significant economic and environmental promise.

4.20 Cropping Systems and Nitrification Inhibitor Efficacy

In wheat–maize cropping systems, DCD and DMPP application led to annual N_2O emission reductions of 35% and 38%. Besides reducing nitrate levels, these inhibitors increased soil inorganic nitrogen, dissolved organic carbon, and crop yield. However, in the Indo-Gangetic plains of South Asia, no-till practices while beneficial for CO_2 reduction may paradoxically raise N_2O emissions. Under such conditions, inhibitors like SBT-butanoate and SBT-furoate showed marked improvements in N_2O reduction and wheat yield enhancement (Bhatia et al. 2010b).

Use of chlorinated pyridine (CP) on winter wheat resulted in the lowest yield-scaled N_2O emissions (0.15–0.17 kg N_2O-N per ton of grain yield) and the highest overall yield under both no-till and conventional conditions. Nitrapyrin and CP improved nitrogen use efficiency (NUE) by 12.6% and cut annual N_2O emissions by 16.5% in vegetable systems (Gilsanz et al. 2016). Abalos et al. (2014) conducted a meta-analysis evaluating DCD, DMPP (NIs), and NBPT (a urease inhibitor). Results showed a 7.5% increase in crop yield and a 12.9% boost in NUE. Factors like high fertilizer rate, coarse soil texture, and alkaline pH ($\geq$8) enhanced the effectiveness of these inhibitors. However, their cost highlights the need for integrating best management practices to maximize returns.

The effectiveness of NIs is often shaped by environmental conditions. Straw incorporation enhances denitrification, and when paired with PIADIN a mix of pyrazole derivatives it reduces N_2O emissions by 41% (Misselbrook et al. 2014).

4.21 Reduction of Nitrogen Emissions in Pasture Systems

The use of nitrification inhibitors and enhanced efficiency fertilizers (EEFs) in pasture systems can lead to significant reductions in nitrogen (N) emissions, and particularly in the release of nitrous oxide (N_2O), a potent greenhouse gas. A meta-analysis across pasture and cropping systems found that both nitrification inhibitors and enhanced efficiency slow-release fertilizers significantly reduce soil N_2O emissions over multiple seasons. Nitrification inhibitors such as DMPP (3,4-dimethylpyrazole-phosphate) and DCD (dicyandiamide) can reduce N_2O emissions from fertilized pasture by up to 65% compared to standard urea applications. In New Zealand, DMPP (1 kg/ha) and DCD (10 kg/ha) reduced N_2O emissions from cow urine by 62–66% and lowered nitrate leaching in Waikato Horotiu soil significantly. A similar study in England showed DCD reducing emissions by 39% (ammonium nitrate), 69% (urea), and 70% (cattle urine) across 14 experimental sites (Misselbrook et al. 2014). Biological nitrification inhibitors released by certain pasture species (e.g., Brachiaria grasses) can reduce N_2O emissions by 60–90%. in swards where these plants are prevalent, giving a direct N_2O mitigation effect in grazing systems (Saud et al. 2022).

4.22 Role in Nitrification Inhibitor in Sustainable Bioenergy Crop Production

Enhanced efficiency fertilizers (EEFs) are key to lessening the environmental footprint of sustainable energy production, in particular through their ability to reducen-itrous oxide (N_2O) emissions from growing biomass. While bioenergy crops, such as sugarcane, are seen as eco-friendly alternatives, their cultivation can paradoxically boost N_2O release due to traditional nitrogen fertilization methods. EEFs, including those with nitrification inhibitors like dicyandiamide (DCD) and 3,4-dimethylpyrazole phosphate (DMPP), significantly curb these emissions. They do this by slowing down the conversion of ammonium to nitrate in the soil, thereby keeping nitrogen in a more stable form. This not only reduces N_2O escape but also lessens nitrate leaching, ultimately improving nitrogen use efficiency and bolstering the overall climate advantages of bioenergy. By minimizing N_2O losses, EEFs help ensure that the growth of sustainable bioenergy generation genuinely contributes to a future with reduced carbon release into the atmosphere.

4.23 Acetylene for Nitrification Inhibition

Acetylene (C_2H_2) serves as a potent inhibitor of nitrification, primarily by targeting the ammonia monooxygenase (AMO) enzyme. This enzyme, which is found in aerobic ammonia-oxidizing bacteria (AOB) and archaea (AOA), is crucial for the initial and rate-limiting step in which ammonium is converted to nitrite. Because of this specific and irreversible inhibition, acetylene proves invaluable in nitrogen cycle research, as it allows scientists to distinguish between nitrification and denitrification and to quantify nitrogen fixation through the acetylene reduction assay. In agriculture, the controlled release of acetylene, often from formulations like coated calcium carbide (CCC), shows promise in delaying nitrification in soils (Malla et al. 2005). This delay can enhance the availability of ammonium for plants, minimize nitrogen losses from nitrate leaching and nitrous oxide emissions, and potentially boost crop yields. However, the effectiveness of acetylene usage can vary with soil type and environmental conditions, and practical challenges remain in its application and in the maintenance ofoptimal concentrations in the field.Acetylene (C_2H_2) is capable of completely halting the process at just 0.1% (v/v). Due to its bacteriostatic properties, it effectively targets autotrophic nitrifiers in acidic to alkaline soils. (Malla et al. 2005).

4.24 Nitrification Inhibitors and Ammonia Volatilization

Nitrification inhibitors (NIs) are primarily used to retain nitrogen in the form of ammonium (NH_4^+) in the form in soil, allowing more time for plant uptake and reducing nitrous oxide (N_2O) production. However, prolonged NH_4^+ retention may unintentionally increase ammonia (NH_3) volatilization. Soil properties such as high pH, low cation exchange capacity (CEC), and elevated sand content can exacerbate NH_3 emissions (Lam et al. 2017). Both NIs and urease inhibitors (UIs) have independently shown success in minimizing nitrogen losses by reducing nitrate (NO_3^-) leaching and gaseous NH_3 emissions, while improving nitrogen use efficiency (NUE). Despite their combined potential—termed "double inhibitors" (DIs)—few studies have evaluated their synergistic effects.

Soares et al. (2012) examined NH_3 volatilization using volatilization chambers with acidic red latosol soil at 60% water retention capacity. Prilled urea (UR) was treated with DCD (NI) at 5–10% of urea weight and/or NBPT (UI) at 1060 mg/kg. NBPT reduced NH_3 volatilization by 54–78% and delayed peak emissions in comparison to untreated urea. However, DCD alone had minimal impact and, when combined with NBPT, appeared to increase volatilization due to prolonged NH_4^+ retention and elevated soil pH. This suggests that the effects of interactions between NIs and UIs can sometimes negate the benefits of NH_3 reduction offered by UIs alone.

Nonetheless, with appropriate management including irrigation scheduling and consideration of climatic conditions combined NBPT and DCD applications have successfully reduced N_2O emissions in maize cultivation by 24–43% in Spain and 46.8% in China. In support of these results, a meta-analysis by Lamet al. (2017) found that combined use of NIs and UIs was more effective in reducing NH_3 emissions than NIs alone. Optimal performance was observed when DIs were applied five days before urine deposition in pastoral systems, especially during cooler autumnal periods.

Although nitrification inhibitors are effective in reducing N_2O emissions and nitrate leaching, emerging evidence suggests that they may inadvertently contribute to increased NH_3 volatilization, thereby indirectly fueling further N_2O production (Lam et al. 2017). Fertilizer application timing also plays a critical role in NH_3 loss, which varies seasonally based on soil moisture, rainfall, and temperature conditions.

A study conducted in Northern Germany (Ni et al. 2014). revealed that a blend of DCD and 1H-1,2,4-triazole (a NI) increased NH_3 volatilization, while N-(2-nitrophenyl) phosphoric triamide (2-NPT), a UI, decreased it by 26–83% These findings underscore the importance of precise timing in the application of urease inhibitors in order to effectively mitigate indirect sources of N_2O emissions through NH_3 control.

4.25 The Role of Double Inhibitors in Balancing Efficiency and Volatilization Losses

While nitrification inhibitors are highly effective in reducing nitrous oxide (N_2O) emissions and nitrate (NO_2^-) leaching, some studies indicate that their use may indirectly elevate ammonia (NH_3) volatilization an additional source of N_2O emissions (Lam et al. 2017). The timing of fertilizer application plays a key role in this process, as seasonal changes in soil moisture, rainfall, and temperature influence NH_3 loss percentages. NIs are designed to maintain nitrogen in the form of ammonium (NH_4^+) form for extended periods, thereby enhancing plant uptake and limiting N_2O production. However, this prolonged retention may increase NH_3 volatilization, particularly in soils with high pH, low cation exchange capacity (CEC), or high sand content (Kim et al. 2012). Although both NIs and UIs independently improve nitrogen use efficiency and minimize gaseous losses, few studies have assessed their combined impactas double inhibitors. In combining both urease and nitrification inhibitors, the latter mark a significant step forward in curbing various forms of nitrogen loss from agricultural systems. Urease inhibitors work by targeting the initial breakdown of urea into ammonium, thereby cutting down on ammonia volatilization—a gaseous loss that often takes place when urea-based fertilizers are applied to the soil surface. At the same time, nitrification inhibitors prevent the subsequent change of ammonium into nitrate, which is highly prone to leaching into groundwater and undergoing denitrification, in a process that releases potent greenhouse gases like nitrous oxide (N_2O).

By tackling these distinct yet connected pathways of nitrogen loss, double inhibitors offer comprehensive protection. This keeps more plant-available nitrogen in the soil for longer, boosting nitrogen use efficiency, improving crop yields, and substantially reducing the environmental impact of fertilization. Using a double inhibitor approach was shown to decrease both NH_3 volatilization and N_2O emissions from urea-treated sources in a study by Zaman and Blennerhassett (2010). Soares et al. (2015) evaluated the co-application of NBPT (UI) and DCD (NI) on prilled urea (UR) in acidic red latosol soil with 60% water retention. NBPT alone reduced NH_3 volatilization by 54–78% and delayed peak emissions. DCD, however, had limited effect and even increased volatilization when combined with NBPT due to extended NH_4^+ presence and elevated soil pH. Despite these drawbacks, when paired with proper irrigation and favorable climate conditions, NBPT + DCD has demonstrated reductions in N_2O emissions by 24–43% and 46.8% in maize fields in Spain and China, respectively (Ding et al. 2011; Sanz-Cobena et al. 2012). Field trials involving biochar and urea treated with hydroquinone (UI) and DCD (NI) have shown that DIs can reduce NH_3 losses by 19.8% while significantly improving rice yields (He et al. 2018). A similar study in wheat by Thapa and Chatterjee (2017) reported NH_3 and N_2O emission reductions of 34% and 43% using DIs.

Combining Agrotain (UI) with cow urine reduced NH_3 volatilization by 51%, though N_2O emissions remained unaffected. In contrast, pairing DCD with cow urine reduced N_2O emissions by 35.67% but raised NH_3 losses by 27.33%. When both DCD and Agrotain were used together, NH_3 and N_2O emissions were reduced by 33.67% and 44%, respectively. This dual inhibition approach also improved pasture dry matter yield by 9.67% and nitrogen uptake by 18.33%. These effects are attributed to the complementary actions of NBPT and DCD, with the former limiting NH_4^+ availability and the latter increasing, NH_4^+ retention in soil. Meta-analysis by Kim et al. (2012) confirmed the benefits of DIs in mitigating NH_3 emissions, especially in pastoral systems during the cooler autumn months when applied five days before urine deposition in pastoral systems. These findings emphasize the importance of using UIs with careful timing to mitigate secondary sources of N_2O emissions.

4.26 Interaction Between NIs and UIs: Balancing Efficiency and Volatilization

Ammonium thiosulfate (ATS) has shown potential as both a nitrification and urease inhibitor in incubation studies conducted at 25 °C (Goos 1985). When 1 mmol sulfur (as sodium thiosulfate) was added per kilogram of soil alongside 10 mmol nitrogen (as ATS), nitrification was inhibited by 55–80%. Increasing sulfur concentration to 10 mmol/kg enhanced inhibition to 82–91%. Additionally, adding 1% and 10% ATS to a urea–ammonium nitrate (UAN) solution reduced soil urease activity (Margon et al. 2015). ATS significantly suppressed nitrate formation, with the degree of inhibition varying in different soil type. This effect helps keep nitrogen in the stable ammonium form for longer periods, reducing N loss through leaching and denitrification (Margon et al. 2015).

4.27 Limus-Coated Urea and Gaseous N Losses

Paul et al. (2024) and Bhatia et al. (2023) reported that Limus-coated urea was effective in reducing various nitrogen losses from agricultural systems, ultimately leading to more efficient nitrogen use and a lower environmental footprint in rice production. Limus coated urea demonstrates a greater reduction of nitrous oxide (N_2O) emissions compared to conventional prilled urea due to its formulation as a dual inhibitor—combining two urease inhibitors, NBPT (N-(n-butyl) thiophosphorictriamide) and NPPT (N-(n-propyl) thiophosphorictriamide). These inhibitors work synergistically to slow the hydrolysis of urea in the soil, resulting in a delayed and more gradual release of ammonium (NH_4^+). By decreasing the rapid spike in soil ammonium concentration, Limus-coated urea reduces the substrate available for nitrifying bacteria, which in turn slows the production of nitrate (NO_3^-) the key precursor for N_2O formation during the nitrification and denitrification processes. Additionally, this delayed conversion limits not only ammonia volatilization but also the subsequent nitrification and denitrification steps responsible for N_2O emissions.

Chakrabarti et al. (2024) found thatLimus impacted nitrous oxide (N_2O) emissions in wheat systems under future climate scenarios involving elevated CO_2 and temperature. Limus was particularly effective under conditions of elevated temperature and CO_2, when nitrogen losses tend to be higher. By maintaining nitrogen in less mobile and less reactive forms for longer, Limus coated urea increased nitrogen uptake efficiency for plants and reduced gaseous nitrogen losses to the environment. Limus acted like a dual inhibitor disrupting both urease and, to a lesser extent, nitrification processes, making it significantly more effective at curbing N_2O emissions than regular prilled urea. Compared to conventional prilled urea, Limus-coated urea reduced N_2O–N emissions by 14% under combined elevated CO_2 and temperature conditions. Additionally, Limus application lowered ammonia (NH_3–N) emissions by 35.7–36.8% and improved grain nitrogen content by 9%, while enhancing nitrogen recovery efficiency in wheat. Limus effectively decreased both gaseous nitrogen losses and greenhouse gas intensity, making it a valuable tool for improving nitrogen use efficiency and climate resilience in cereal production as global temperatures and atmospheric CO_2 concentrations rise.

4.28 Challenges to the Widespread Adoption of Enhanced Efficiency Fertilizers

Enhanced Efficiency Fertilizers (EEFs), including Controlled-Release Fertilizers (CRFs) and Slow-Release Fertilizers (SRFs), face several challenges that limit their widespread adoption by farmers, despite their environmental benefits, These include.

High Cost: The most significant barrier is the higher material cost compared to conventional fertilizers. While EEFs can potentially reduce total fertilizer use and labor, the initial investment is often a deterrent for farmers and especially to those with limited financial resources.

"Tailing" Effect and Unpredictable Release: For some CRFs, a "tailing" effect can occur, in which nutrients continue to be released beyond the optimal crop uptake period, and so reducing economic benefits. SRFs, which rely on microbial degradation, can have unpredictable release rates influenced by variable soil and climate conditions (e.g., temperature, moisture, pH, microbial activity), making it difficult to match nutrient supply precisely with crop demand.

Lack of Standardized Testing and Correlation: There is a lack of standardized methods to reliably determine the nutrient release rate from CRFs, which is often coupled with and often a disconnect between laboratory data and actual release rates in field applications. This makes it challenging for farmers to trust and effectively utilize these products.

Environmental Concerns with Coatings: Some polymer-coated CRFs use synthetic materials that are difficult to degrade in soil and can accumulate over time, potentially contributing to microplastic contamination and other forms of pollution. Sulfur-coated urea (SCU) can also increase soil acidity when used in large quantities.

Logistical and Application Considerations: While CRFs can reduce the number of applications necessary, the specific requirements for different EEF types (e.g., timing, crop cycle length, soil fumigants) add complexity to farm management. Farmers may also lack sufficient knowledge or guidance on the proper use and benefits of these specialized fertilizers.

Farmer Perceptions and Practices: Farmers may have a lack of confidence in new fertilizer technologies, preferring established conventional practices. Socioeconomic factors like farm income and access to credit also play a role in the willingness to invest in more expensive, albeit potentially more efficient, options.

4.29 Pathways Ahead

Promoting research and development into more environmentally friendly and biodegradable coating materials for CRFs is needed to address concerns about microplastic accumulation.

For EEFs to be more widely accepted by farmers, a multi-faceted approach addressing their current limitations and enhancing perceived benefits is necessary.

Convincing demonstrations of clear and obvious economic benefits are necessary, together with extensive outreach support and targeted government incentives for farmers.

Field-based Research and Demonstrations: Large demonstrations comparing EEFs with conventional fertilizers and highlighting the return on investment are required. Conducting more localized, long-term field trials across diverse soil types, climates, and cropping systems may provide tangible evidence of the' benefits of EEFs, such as increased yields, improved nutrient use efficiency, and reduced labor costs from fewer applications. The dissemination of clear cost–benefit analyses is essential to support informed decision-making around enhanced-efficiency fertilizers (EEFs). These analyses should not only highlight the higher upfront costs associated with EEFs but also provide a comprehensive view of their long-term economic advantages.

Extension outreach: It is important to provide comprehensive training and extension services to farmers on the appropriate selection, application rates, timing, and integration of EEFs into existing nutrient management plans. This includes offering guidance on how to optimize EEF use for specific crops, soil types, and environmental conditions.

Peer-to-Peer Learning: Opportunities for farmers to share their experiences and successes with EEFs, are necessary to foster a sense of community and trust in the technology.

Government Incentives and policy framework: Here the implementation of government or industry programs that offer cost-share opportunities or financial incentives to offset the higher initial cost of EEFs is required These programs should be aimed at encouraging farmers to try out these new techniques and subsequently adopt them. Governments can play a crucial role by developing clear regulatory frameworks and standards for EEFs, ensuring their safety and efficacy. Policies that reward efficient fertilizer use or target support away from blanket subsidies on conventional fertilizers towards more targeted support for EEFs could also drive up adoption.

Decision Support Tools: Develop user-friendly digital tools (such as apps or software) that can help farmers select the right EEF product, calculate optimal application rates, and predict nutrient availability based on real-time field data.

By focusing on these areas, stakeholders can build farmer confidence, demonstrate tangible benefits, and create a supportive environment for the wider acceptance and adoption of EEFs.

Chapter 5
Climate Smart Nitrogen Management

5.1 N as an Essential Nutrient

Nitrogen (N) is one of the most vital nutrients for plant growth and a cornerstone of global food production systems (Islam et al. 2024a; Al-Amin et al. 2024a). It is a double-edged sword for global food security and environmental sustainability. As of now, the world's population has reached approximately 8.2 billion and is projected to climb to 10.4 billion by the year 2100 (UN 2024). This escalating demographic pressure poses serious challenges for both agriculture and the environment (Rahman et al. 2022a). To meet the growing demand for food, agriculture must contend with multiple stressors, including soil degradation, increasing climate extremes, global warming, and reduced crop yields associated with climate change (Islam et al. 2018; Rahman and Billah 2024). Addressing these issues demands intensification of agricultural output primarily through high-yielding crop varieties, widespread use of inorganic fertilizers, and adoption of modern technologies. These will be done within the constraints of limited arable land and diminishing water resources. It is estimated that to feed the global population in 2050, food production will need to increase by 60–110% compared to levels recorded between 2005 and 2007 (Pugh et al. 2016; Rahman et al. 2020; Roy et al. 2021). However, such high-input agricultural systems come at a cost: declining soil health, stagnating crop yields, disrupted ecosystem functions, and rising environmental pollution. Nitrogen plays a central role in this scenario. On one hand, it is indispensable for producing food, feed, and fiber making it a blessing to humanity. On the other, excessive reactive nitrogen (Nr) contributes profoundly to ecosystem degradation, earning its place as a major contributor to global pollution. The dual nature of nitrogen as a driver of productivity and a source of environmental harm demands careful management to balance the needs of agricultural sustainability with planetary health (Fig. 5.1).

In this context, technological innovations and improved agronomic practices offer promising avenues to enhance nitrogen use efficiency (NUE) while reducing environmental burdens. The development of an Android-based mobile application supports

U. C. Kulshrestha et al., *Atmosphere-Biosphere Interactions of Reactive Nitrogen in South Asia*, Springer Briefs in Interdisciplinary Geosciences – Asia-Pacific,
https://doi.org/10.1007/978-3-032-17194-8_5

Fig. 5.1 Essentiality and challenges of nitrogen fertilizer in agriculture and environment (Rahman et al. 2024)

precise, data-driven fertilizer decisions, minimizing nitrogen losses across rice-based systems. Similarly, biochar application, deep placement of urea super granules (USG), and leaf color chart (LCC)-guided nitrogen management have demonstrated substantial gains in rice yield, NUE, and farmer income. Biochar enhances long-term soil health, deep placement of USG ensures efficient nutrient uptake and higher economic returns, and LCC-based application synchronizes N supply with crop demand. However, despite their scientific merits and climate-smart potential, adoption remains low due to limited farmer awareness. Scaling up training, demonstrations, and digital advisory services is therefore crucial for mainstreaming these environmentally friendly practices and aligning nitrogen management with the broader goals of sustainable agricultural intensification and planetary health. These practices are classified as climate-smart because they boost productivity and farm profitability, reduce environmental footprints and support climate mitigation, strengthen resilience to climate variability, enable large-scale adoption through capacity building and promote global collaboration for sustainable agriculture. Together, they embody the integrated approach required for climate-smart agriculture. Therefore, the objectives of this chapter are to provide an overview of the climate-smart nitrogen management practices, create awareness on such practices among the concerned stakeholders and provide recommendation for policy makers. Climate-smart nitrogen management enhances crop productivity and farmer profitability while simultaneously reducing nitrogen losses, improving soil health, and mitigating environmental pollution and greenhouse gas emissions.

5.2 N Fertilizer Consumption

5.2.1 Nitrogen Deficiency and Fertilizer Recovery Challenges

Nitrogen (N) is universally deficient across most agricultural production environments, making its external application essential for achieving desired crop yields. However, the efficiency of nitrogen uptake remains a major concern: only 20–40% of applied nitrogen fertilizer is typically recovered in harvested crops, while a substantial 60–80% remains unused and lost to the environment (Ladha et al. 2003; Rahman 2013; Alam et al. 2023; Rahman et al. 2024). Adding to the challenge, rapid industrialization, expanding urbanization, and climate change are contributing to a continuous decline in cultivable land area (Hasan et al. 2019a; Rahman 2024), prompting farmers to increase inorganic fertilizer input per unit of land to sustain food production.

5.2.2 Global Trends in Fertilizer Application

Data illustrated in Fig. 5.2 tracks global usage trends of nitrogen (N), phosphorus (P), and potassium (K) fertilizers from 1970 to 2021, showing a steep rise in nutrient inputs over time. Figure 5.3 further highlights regional and country-level shifts in nitrogen fertilizer application between 2000 and 2022. During this period, nitrogen use expanded significantly across most regions and countries, with the notable exceptions of Sri Lanka and Vietnam. This expansion mirrors the diminishing inherent nutrient-supplying capacity of agricultural soils, compelling greater reliance on synthetic nitrogen fertilizers. As shown in Fig. 5.4, such fertilizers now account for more than half of the world's food grain output (Tandon and Tiwari 2007; Jiao et al. 2016).

5.2.3 South Asia: A Global Hotspot for Nitrogen Use and Pollution

Among global regions, South Asia (SA) stands out as a concentrated hotspot for nitrogen fertilizer use and associated pollution. Despite comprising less than 5% of

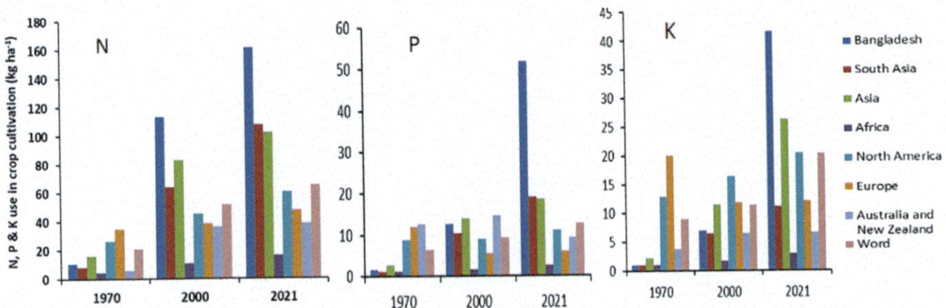

Fig. 5.2 Nitrogen (N), phosphorus (P) and potassium (K) application in agriculture in different regions of the world from 1970 to 2021 (FAOSTAT 2023; Rahman 2024)

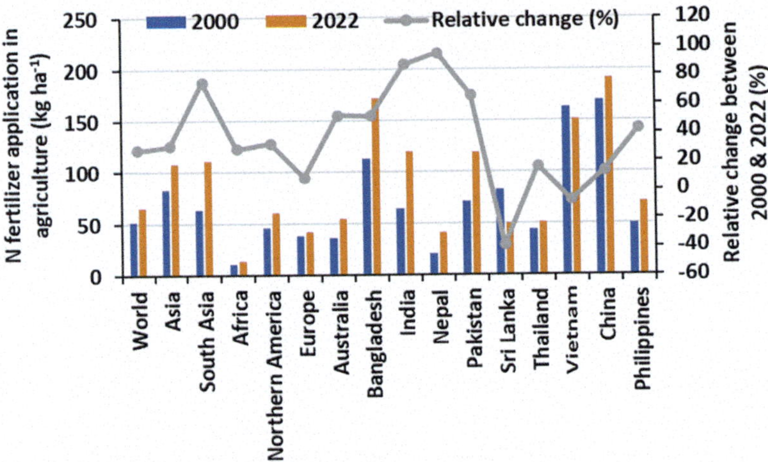

Fig. 5.3 Nitrogen fertilizer application and its relative changes in different regions and countries of the world during 2000 to 2022 (*Data source* FAOSTAT 2024)

the global land mass and just 14% of the world's arable land, SA supports over 25% of the global population and nearly 45% of Asia's population. The region consumes approximately 58% of the total nitrogen fertilizer used in agriculture worldwide, as depicted in Fig. 5.5. Between 2000 and 2022, nitrogen fertilizer uses in SA rose by 75% (Fig. 5.3). Due to persistently low nitrogen use efficiency (NUE), a substantial portion of applied reactive nitrogen (Nr) remains unutilized, posing serious environmental risks both regionally and globally (Billah et al. 2024).

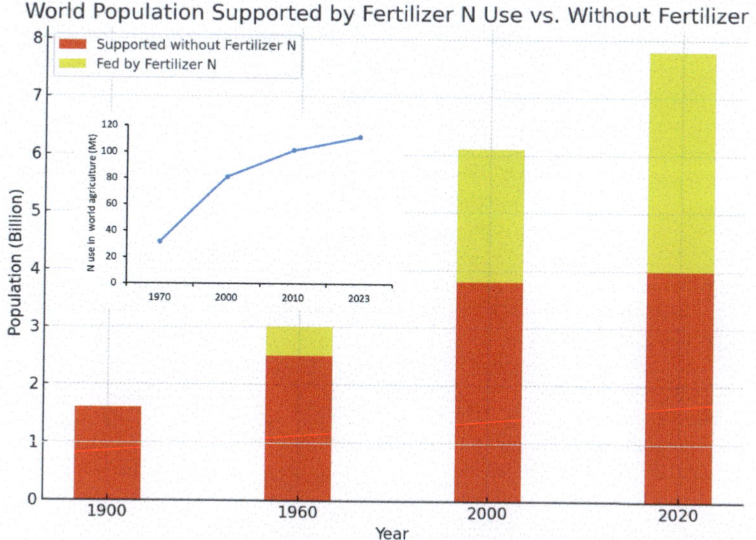

Fig. 5.4 Soil capacity is declining to supply nutrients to plants, while need for N fertilizer is increasing for crop production (*Data source* IFADATA 2020; Singh and Craswell 2021; FAOSTAT 2025)

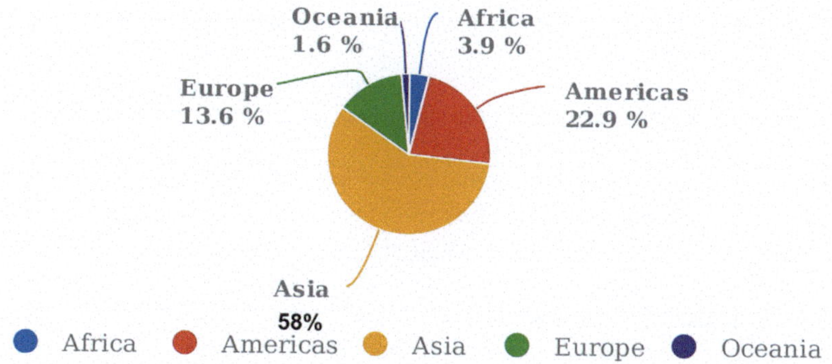

Fig. 5.5 Nitrogen use in agriculture in 2020: share by region (FAOSTAT 2023)

5.3 Nitrogen Management for Sustainable Agriculture and Environmental Protection

Excessive reactive nitrogen (Nr) including ammonia, nitrate, and nitrous oxide poses serious threats to soil, water, and air quality and contributes significantly to global climate change (Fig. 5.6). Imbalanced or excessive application of nitrogen fertilizers, particularly in intensively cultivated systems, has become a pressing issue in many countries. This practice accelerates soil degradation, increases nutrient losses, and

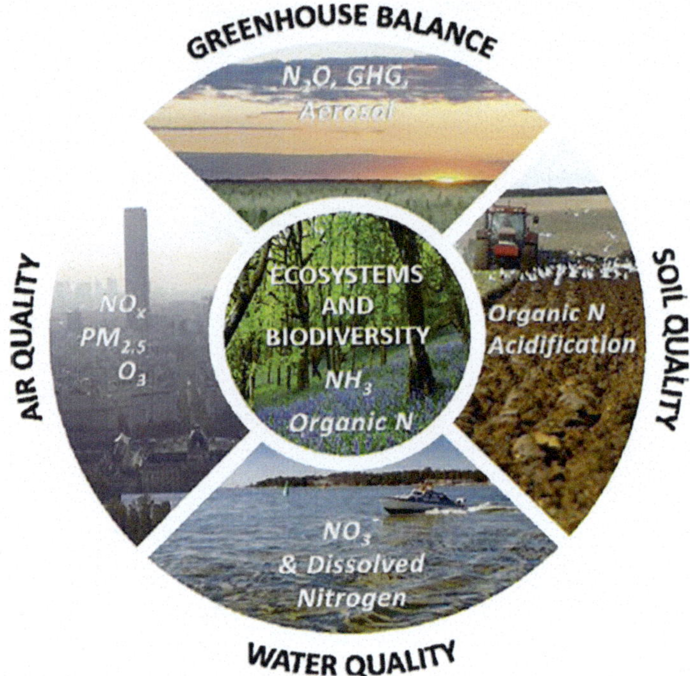

Fig. 5.6 Key threats of nitrogen pollution in the environment (Sutton et al. 2013)

leads to widespread environmental contamination (Rahman 2014; Islam et al. 2018; Rahman et al. 2021; Hasnat et al. 2022).

Meanwhile, the availability of arable land is steadily shrinking due to industrial expansion, rapid urbanization, and climate-induced stress. In Bangladesh alone, the annual loss of cultivable land is estimated at 87,000 hectares (Hasan et al. 2019b; Rahman et al. 2021). In such a context, achieving optimal crop yields from limited land and water resources requires a carefully balanced supply of nitrogen fertilizer used alongside other essential nutrients to sustain productivity (Shashi et al. 2018; Alam et al. 2023; Anik et al. 2023).

For long-term agricultural sustainability and environmental stewardship, it is critical to adopt best practices in soil, crop, and fertilizer management. Nutrient losses from soil particularly in sensitive ecosystems like wetlands are a major contributor to greenhouse gas emissions (Alam et al. 2014; Bhatia et al. 2023). Moreover, excessive application of fertilizers not only undermines ecological integrity but also inflates production costs, especially under the burden of rising global energy prices (Rahman et al. 2022b; Islam et al. 2024b).

5.3.1 Nitrogen Use Efficiency (NUE)

A key metric for addressing these challenges is nitrogen use efficiency (NUE), defined as the proportion of applied nitrogen that is successfully taken up and utilized by the crop for biomass and grain production. Low NUE means that a large percentage of nitrogen remains unused, escaping into the environment through volatilization, leaching, or denitrification. Improving NUE is essential for reducing nitrogen losses, minimizing environmental harm, and enhancing crop yields without increasing fertilizer input.

Hence, there is an urgent need to identify practical, region-specific strategies to boost NUEparticularly in South Asia and other vulnerable agricultural zones. Enhancing NUE not only ensures food security for a growing population but also helps in mitigating the ecological footprint of modern agriculture.

5.4 Addressing the Nitrogen Challenge in South Asia: A Collaborative Effort

South Asia (SA) is among the most heavily impacted regions globally when it comes to nitrogen (N) pollution. Few places on Earth experience levels of reactive nitrogen accumulation as severe as this region, making it a focal point for global environmental concern. The consequences are far-reaching which threatening biodiversity, contributing to climate change, and undermining public health and ecological integrity. Urgent action is needed to reverse this trend and transition toward more sustainable nitrogen management practices.

To respond to this challenge, a multidisciplinary coalition was established under the South Asian Nitrogen Hub (SANH), consisting of 32 leading research institutions and partner organizations from both South Asia and the United Kingdom. SANH scientists have focused on optimizing nitrogen use in agriculture by improving fertilizer application practices, harnessing natural nitrogen fixation processes, and promoting the recycling of organic waste materials such as manure and urine within a circular economy framework. Their research highlights practical pathways to cleaner, more efficient, and more profitable farming systems tailored to South Asian conditions.

The region holds critical importance in the global nitrogen cycle. With a population exceeding 1.7 billion and growing at an annual rate of 1.7%, South Asia's demographic pressures are expected to double by 2050. Correspondingly, agricultural nitrogen inputs are projected to double as well, positioning SA as the top global consumer of nitrogen fertilizers. Compounding this issue is the region's low nitrogen use efficiency (NUE), coupled with extensive government subsidies that incentivize over application. For instance, India allocates approximately £6 billion annually for fertilizer subsidies (SANH 2020), while Bangladesh contributes nearly £2 million (The Business Standard 2022).

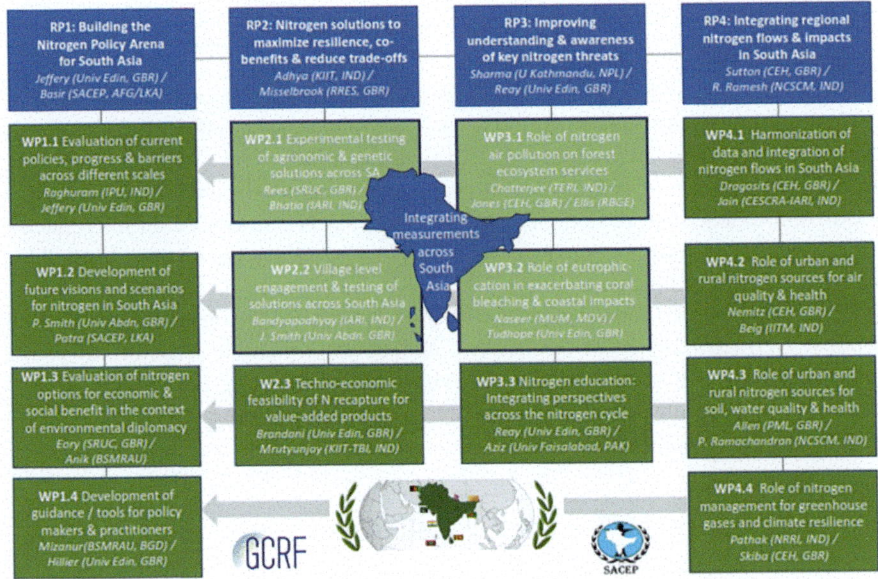

Fig. 5.7 Research structure of the South Asian Nitrogen Hub (SANH) comprised of different work packages (WPs)

These financial commitments underscore the importance of adopting more strategic nitrogen management approaches. The collaborative efforts between South Asian and UK scientists are geared toward identifying innovative solutions that can be translated into field-level practices. However, to make these solutions truly effective, active policy support and governmental engagement are essential. The South Asia Cooperative Environment Programme (SACEP) plays a pivotal role in bridging scientific research with policymaking, ensuring that SANH's findings are implemented across the region. Details of SANH's research structure and operational activities are illustrated in Fig. 5.7.

The key goal of the hub is to establish a research partnership between the UK and South Asia, to promote sustainable N management in the SA region. The vision is to lead the way in managing N effectively, driving positive change towards achieving the Sustainable Development Goals (SDGs) worldwide. The objectives of the hub are to:

- Reduce poverty and hunger by promoting sustainable farming methods that produce more food with fewer chemicals,
- Identify best N management options that reduce N losses, increase NUE, and ensure sustainable agriculture and environment,
- Take action on climate change by cutting greenhouse gas emissions and promoting climate-resilient practices in food production,

- Educate and aware different stakeholders about N challenges and its impacts on soil and environment, and motivate and ensure stakeholders' engagement on sustainable N management,
- Liaison with the government for implication of supportive policy and formulation of new policies in favour of technology dissemination and adoption,
- Protect water resources by reducing pollution and addressing threats like coral bleaching, and
- Preserve biodiversity by understanding and mitigating the impact of N pollution on ecosystems, especially in the Himalayan region.

5.5 Stakeholders' Engagement in N Management

This chapter of the monograph is a compendium of SANH research findings and stakeholders' engagement activities at the Gazipur Agricultural University (GAU), Bangladesh (Former Bangabandhu Sheikh Mujibur Rahman Agricultural University i.e., BSMRAU). GAU completed a good number of research and stakeholders' engagement activities for sustainable N management. Major findings/deliverables achieved from different perspective of N management are briefly presented below.

5.5.1 Current Policies and Progress for N Management

The key output of the fertilizer policy related issue is the Bangladesh Nitrogen Policy Report: Scientific Evidence, Current Initiatives, and Policy Landscape (Shifa et al. 2022). This is the first report to address N issues from a policy perspective. Previously, environmental policy discussions entered on carbon, with little attention to N. This report has sparked interest in N among socio-economic researchers and policymakers. On June 22, 2022, in Dhaka, the South Asia Cooperative Environment Programme (SACEP) and GAU jointly convened a meeting where scientists from all South Asian partner countries presented the regional report on N pollution policies. National reports from Bangladesh, Sri Lanka, Maldives, and Nepal were also released, documenting policies to address N pollution's threats to the environment, climate change, food security, human health, and the economy. In the meeting, the regional and Bangladesh nitrogen policy reports were released stressing the need for cooperation on sustainable N management in agriculture. Stronger scientist-policy maker collaboration on N challenges is required for sustainable agriculture and environment.

Key Contributions of the N Policy Document (Shifa et al. 2022)

(a) **Nitrogen Policy Database**: A comprehensive South Asia nitrogen policy database has been created evaluating 966 policies by sector, sink, type, and quality, revealing gaps and opportunities in nitrogen management. (Bangladesh: Overview of 187 policies specific to Bangladesh).
(b) **Sector Focus**: About 35% of the policies targeted exclusively agriculture, while only 5% addressed multiple sectors and environmental sinks.
(c) **Emission Drivers**: Agriculture, industry, and transport are major Nr sources in Bangladesh. Urea use increased by 11.7% since 2002, while agriculture produced 85% of NH_3, industries emitted approximately 43% of NOx, and transport emissions have surged since 2000.
(d) **Policy Direction**: Findings revealed that about 44% of policies positively impacted Nr management, but increased in Nr compounds signal a need for wider-scale policy efforts for sustainable nitrogen management aligning with UN SDGs.

5.5.2 *Nitrogen Options for Social and Economic Benefits*

The N fertilizer issues has been successfully highlighted in the 8th Five Year Plan, marking the first-time concerns about N use in agriculture were included in a Bangladeshi government policy document. The report raised concerns over excessive urea use, which makes up about 75% of all fertilizer nutrients used in the country, potentially harming soil health and contributing to water pollution. It recommended balancing urea and non-urea fertilizer subsidies to promote more balanced fertilizer use. These findings have positioned GAU as a pioneer in N policy research, leading to regular communication between peer scientists and the GAU team, which has strengthened collaboration and awareness on N pollution issues. However, awareness among policymakers remains limited, hindering the implementation of coordinated, impactful nitrogen management policies. A social survey has been conducted involving 720 farmers from 24 villages across 24 unions in 12 sub-districts within 6 districts of 2 divisions in Bangladesh. The survey aimed to understand how farmers used N for both crops and livestock, providing insights into their decision-making processes.

5.5.3 *Key Contributions of Identifying N Management Options*

(a) **Addressing Soil Health in Bangladesh**: Focused on the critical issue of soil degradation, with limited prior research on sustainable soil management (SSM) adoption in Bangladesh.

(b) **Comprehensive Classification**: Identified and categorized commonly used 15 SSM practices across five key management areas viz., crop, residue, nutrient, tillage, and water management.
(c) **Determinant Analysis**: Used advanced econometric models to identify factors influencing SSM adoption, offering insights into socio-economic, demographic, and geographic drivers.

5.5.4 Nitrogen Management Tools and Guidance

Under this activity a free-to-use N app has been created for the farmers in decision making on fertiliser and manure application more efficiently across the South Asian region. At first, were viewed, characterized and identified 53 tools and apps for N management from South Asian (30) and other countries (23). Based on the review studies and our research on agronomic testing and finding solutions for N management, four N management tools and apps viz., Soil Test-based N (STBN), Leaf Color Chart-based N (LCCN), SPAD meter-based N (SPADN) and Bangladesh Soil Resource Development Institute (SRDI) App-based N (SRDIAppN) were selected for conducting an experiment and along with a N control treatment to assess their performance. The study was conducted in two successive seasons of Boro (dry season rice with full irrigation) using the variety BRRI dhan29 and Transplanted Aman (T. Aman) (rainfed rice with supplemental irrigation if needed) using the variety Binadhan-7 during 2022.

In case of Boro rice, the findings showed that grain yield was significantly highest in the LCCN and STBN treatments (6.67 t ha^{-1}), followed by SRDIN (6.2 t ha^{-1}), SPADN (5.89 t ha^{-1}) and control (2.98 t ha^{-1}). Agronomic efficiency of N was appeared significantly highest in the LCCN and SPADN treatments (25 kg grain kg^{-1} N applied) compared to all other treatments. The significantly highest physiological efficiency was found in the SPADN (46 kg grain kg^{-1} N uptake), while LCCN (40 kg grain kg^{-1} N uptake) and STBN (38 kg grain kg^{-1} N uptake) treatments were found statistically similar, and the lowest was in the SRDIN treatment. Nitrogen recover efficiency was found the significantly highest in the LCCN (63%) followed by SPADN (53%), STBN (42%) and SRDIAppN (35%). Similar trends of grain yield and N use efficiencies were also observed in case of T. Aman rice. Findings revealed that LCC can be recommended as farmers practice for N fertilizer application to increase rice yield and NUE. The global scientists are in thirst for innovating location specific best management options to raise NUE to a reasonably higher level from the present state. LCC is a good technology in terms of environmental sustainability and economic viability but its adoption rate is low to medium. Even though it is a good technology but not farmers' friendly. Therefore, it was hypothesized that if we can transfer the LCC technology to the android mobile farmers may use it comfortably. Based on the hypothesis GAU in collaboration with the University of Edinburgh developed Android-based N Management App using the LCC technology. The LCC

Fig. 5.8 Validation trial of developed N app at GAU research field

features have been transferred to the Android mobile through image processing technology. It can evaluate greenness levels by comparing images of rice leaves from farmers' fields with LCC reference values and recommend fertilizers for rice. The newly developed N app was validated at the GAU research field (Fig. 5.8).

After on-station validation, the developed App will be validated through national and regional trials. After successful field trials the App will be released for farmers' practice as an N guidance tool for rice cultivation in the South Asian region. It is expected that the N app will facilitate the N management system in rice through mobile phones, optimize N use efficiency and increase rice yield.

Research achievements and findings from N app development were presented in different international and national conferences/seminars. Among which two notable presentations were in Hanoi, Vietnam, and Madrid, Spain (Fig. 5.9). Participants of the conferences were very curious to know about mobile-based N app development and appreciated very much as this tool might reduce N rates for rice cultivation. This ultimately will contribute to reduce GHG emission from rice field and mitigate the negative effects of climate change.

Key Contributions from the N App Development Study

- Results from review analysis and experiment conduction revealed that deep placement urea super granule (USG) and leaf color chart (LCC) are efficient for N management, but their adoption rates are low,
- Mobile-based N app has been developed based on LCC using image processing technology,

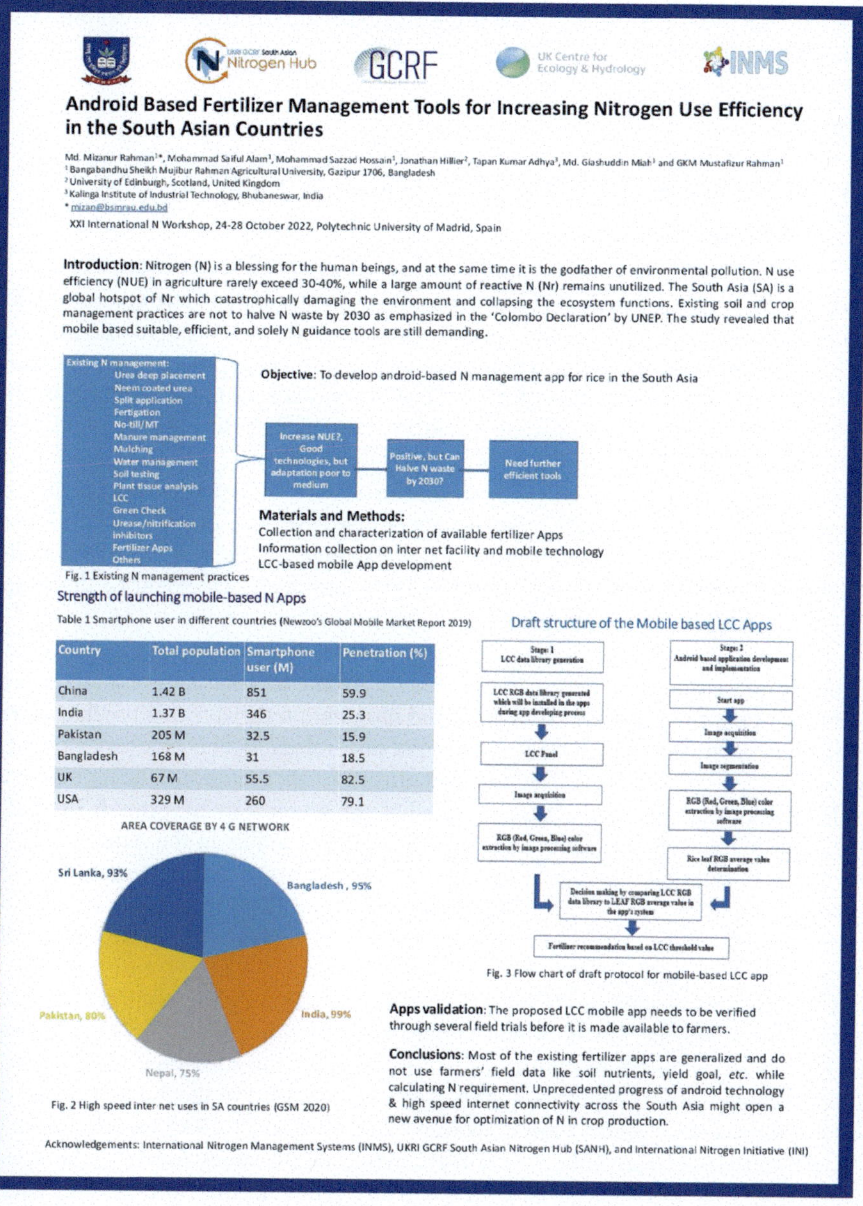

Fig. 5.9 Poster on N App development presented in the XXI International N Workshop held in Madrid, Spain during 24–28 October 2022

- Validation trial of the N App is ongoing and expected to promote its adoption and further increase NUE.

5.6 Experimental Testing and Agronomic Solutions of N Management

Under the activity we conducted on-station experiments using some common South Asian practices and some country specific innovative approaches to identify the best options to reduce N pollution without compromising crop yield and food quality, and other negative impacts (Fig. 5.10). The study was conducted in four consecutive seasons of Boro and Transplanted Aman (T. Aman) rice during 2021 and 2022. Boro is a dry season rice grown with full irrigation, while T. Aman is grown under rainfed condition with supplementary irrigation if needed. The experiment comprised seven treatments, viz., zero N (control), recommended dose of N (RDN), 125% of RDN (RDN125), 75% of RDN (RDN75), cow dung 2 t ha^{-1} + supplemented N (CDSupN), biochar 2 t ha^{-1} + RDN (BRDN), and deep placement of urea super granules (USG). The recommended N rates of prilled urea were 186 kg ha^{-1} in the Boro and 102 kg ha^{-1} in the T. Aman, while those for USG were 95 kg ha^{-1} in the Boro season and 75 kg ha^{-1}in the T. Aman season. The objectives were quantification of the impacts of organic and inorganic fertilizers and biochar on rice yields, nitrogen use efficiencies (agronomic, physiological and recovery), nitrogen losses through volatilization, leaching and nitrous oxide emission. The highest grain yield of Boro rice was in the BRDN treatment at 6.6 t ha^{-1} in 2021 and 6.7 t ha^{-1} in 2022, while the same treatment also produced higher yields in the Aman season at 4.6 t ha^{-1} in 2021 and 4.8 t ha^{-1} in 2022. Compared with the RDN treatment, grain yield increased by 15% in the BRDN and 8% in the USG treatments in the Boro season, while for the Aman season there was a 17% increase in the BRDN and 9% in the USG.

Leaching losses of NH_4^+–N varied between 1.3 and 9.4 kg ha^{-1} in the Boro season and 0.9 and 5.9 kg ha^{-1} in the Aman season, while leaching of NO_3^-–N ranged from 1.4 to 11.8 kg ha^{-1} in Boro season and 0.7 to 4.4 kg ha^{-1} in Aman season (Table 5.1). Total leaching losses (NH_4^+–N + NO_3^-–N) are provided in the Table 2.1. In Boro season, the average N_2O-N emissions were 2.58, 1.98 and 1.15 kg ha^{-1} in the RDN, BRDN and USG treatments, respectively, while these values for T. Aman season were 1.56, 1.08 and 0.82 kg ha^{-1}, respectively. Thus, in the Boro season, BRDN and USG contributed to reduce N_2O-N emission by 24% and 56%, respectively compared to RDN treatment, while these values for T. Aman season were 31% and 47%, respectively. During the four consecutive rice growing seasons, N leaching, NH_3 volatilization losses and nitrous oxide emission followed the order of RDN125 > RDN > RDN75 > CDSupN > BRDN > USG > control. Compared with the applied N, the N leaching losses were only 3% in the BRDN and 4–5% in the USG treatments which were almost half that of the RDN125 and RDN treatments. In proportion to the

Fig. 5.10 Experiment to identify best N management options at GAU, Bangladesh

applied N, the N volatilization losses were only 8–9% in the BRDN and 5–6% in the USG treatments which were almost half as that in the RDN125 and RDN treatments. Losses by NH_3 volatilization exceeded losses by leaching across all N treatments. Biochar with RDN and USG had the greatest potential to reduce N losses without loss of rice yields. The results showed that, as compared to the RDN treatment, the BRDN treatment increased grain yield and recovery N use efficiency (REN) by 15.25–16.88% and 43.15–50.88%, respectively. Compared to the RDN treatment, the USG treatment increased yield and REN by 6.32–10.26% and 31.96–73.51%, respectively. As a result, applying the recommended N dose with biochar and urea super granule was effective for increasing rice yield, improving fertilizer efficiency, and promoting sustainable agricultural development.

The highest gross return (\$1940 ha^{-1}) in *Boro* rice was found in the BRDN, while the highest net return (\$605 ha^{-1}) and BCR (1.51) were found in the USG treatment. Similar results were also observed in case of *T. Aman* with the highest net return of \$480 ha^{-1} and BCR of 1.57 in the USG. The deep placement of USG has plenty of barriers that are experiencing over 40 years of engagement since 1980s which yet to be solved for its adoption at farmers' level. Urea briquetting is presently done at the village level and such small entrepreneurships are reluctant to produce sufficient amount of USG. Lack of large-scale commercial manufacturer of USG seriously inhibits its adoption at the farm level. Secondly, the less availability of farmers' friendly efficient USG applicators acts as an obstruct of the technology implementation. Conversely, manual application is time consuming which requires more labor and farmers feel backpain as they need to bend down during hand placement of USG. Moreover, negative attitude of fertilizer dealers towards USG discourages farmers to adopt USG. Lack of farmers' awareness about the multiple benefits of USG are also the reasons of low adoption of such an excellent technology. On the other hand, utilization of biochar is not popular because of its higher costs, and lack of understanding about its benefits in agriculture. Favourable government policies and investment from private–public sectors and donor organizations might be helpful to address such challenges. If such challenges are addressed, then biochar urea deep

Table 5.1 Total leaching losses of N in wetland rice under different N management practices during 2021 and 2022 (Islam et al. 2024a)

Treatment	N leaching in 2021 (kg ha^{-1})		N leaching in 2022 (kg ha^{-1})		Annual loss in 2021	Annual loss in 2022	Average loss in Boro	Average loss in Aman
	Boro	Aman	Boro	Aman	kg ha^{-1}		% (Applied N)	
Control	3.3f	1.6d	2.7e	1.6e	4.9f	4.3f	–	–
RDN	14.1bA	6.8bB	12.3bA	6.5bB	20.9bA	18.8bB	7	6
RDN125	21.2aA	10.3aB	18.9aA	9.9aB	31.5aA	28.7aB	9	8
RDN75	8.7cA	3.9cB	7.4cA	3.7cB	12.6c	11.1c	6	5
CDSupN	8.3cA	3.6cB	7.2cA	3.5cB	11.9cdA	10.7cdB	5	5
BRDN	6.7dA	3.2cB	5.5dA	3.1cdB	9.9d	8.6de	3	3
USG	4.7e	2.9c	3.9eA	2.6dB	7.6e	6.6e	5	4
CV	5.38	9.46	6.73	7.63	6.43	6.88	–	–
S.E. (±)	0.51	0.44	0.56	0.34	0.91	0.87	–	–

Different lowercase letters in a column indicate significant differences among the values in each column. Different capital letters in a row between Boro and Aman in 2021, between Boro and Aman in 2022, or between 2021 and 2022 for annual loss indicate significant differences between the two times of measurement

placement technology in agriculture could become a more economically viable and environmentally sustainable, and can be recommended as improved practices for rice-growing farmers.

5.7 Consumption of Biologically Fixed Green Nitrogen and Agricultural Sustainability

Biological N fixation (BNF) has significantly contributed to agriculture by offering a sustainable solution to soil N deficiency. Different legume species exhibit annual N fixation through BNF ranging from 33 to 465 kg N ha^{-1} yr^{-1} (Fig. 5.11). Legume crops transfer N to the soil or neighboring plants in varying quantities, ranging from 0 to 650 kg ha^{-1}. Biofertilizers have shown potential in saving N fertilizer, ranging from 20 to 100 kg per hectare depending on the crop and biofertilizer used. BNF contributes to lowering synthetic N fertilizer usage, improving soil and plants' nutrient content, and restoring soil health. The Fig. 16b depicted that with the advancement of time BNF in the crop land is increasing in the Asian region including Bangladesh and the world as a whole. This is really promising for reduction of inorganic N fertilizers application in agriculture. Utilizing biologically fixed N contributes to the long-term sustainability of agriculture and the environment (Rahman et al. 2024).

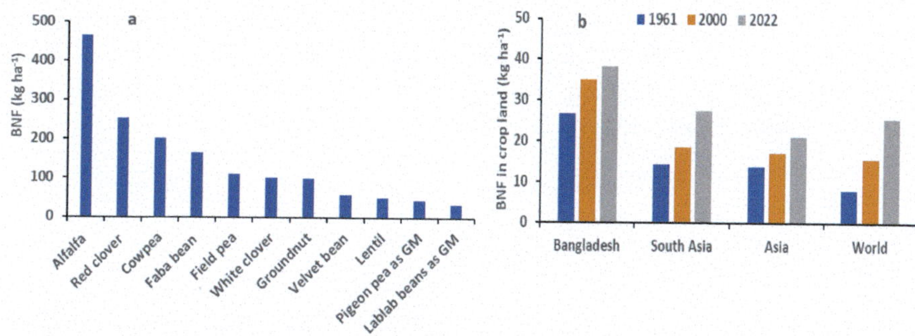

Fig. 5.11 Biological nitrogen fixation: **a** by different plant species (Dakora et al. 1987; Okito et al. 2004; Anglade at el. 2015; Mendonca et al. 2017), and **b** in crop land of South Asia and the world (*Data source* FAOSTAT 2025)

5.8 Soil Health Improvement: Combined Application of Organic and Inorganic Fertilizers in Paddy Cultivation

Organic fertilizers (like compost, manure, green manure) improve soil structure, boost microbial activity, and contribute to long-term fertility. Inorganic fertilizers (such as urea or ammonium sulfate) provide immediate nutrient availability for crop growth but combined application leverages the strengths of both—quick nutrient supply and slow-release soil enrichment. Soil health can be improved through organic inputs, by incorporating organic materials like farmyard manure (FYM), vermicompost, or biochar during land preparation, by using green manure crops like *Sesbania* before transplanting rice.

Nitrogen losses occur through leaching, volatilization, and denitrification which can be stopped by applying urease inhibitors (UIs) and nitrification inhibitors (NIs) with urea to slow nitrogen transformation. Use split application of nitrogen fertilizer—base dose, tillering stage, and panicle initiation also helps in in controlling nitrogen losses. Other options to control nitrogen losses are adopt ion of alternate wetting and drying (AWD) irrigation which improves NUE and reduces CH_4 emissions while minimizing N leaching, and by avoiding flooding immediately after nitrogen application to reduce NH_3 volatilization.

5.8.1 Enhance Nitrogen Use Efficiency (NUE)

NUE is defined as the proportion of applied nitrogen that is absorbed and used by the crop. NUE can be improved by using combine slow-release fertilizers (like polymer-coated urea or sulfur-coated urea) with organic manure; by aligning fertilizer application with crop demand stages—young rice plants require less N than reproductive stages; by applying deep placement of urea super granules in puddled soil to reduce surface losses and by encouraging microbial inoculants (e.g., *Azospirillum*, *Azotobacter*) that fix atmospheric N and improve root uptake.

5.9 Soil and Water: Essentials of Life

South Asia faces intensifying pressure to sustain agriculture in the wake of shrinking cultivable land, declining water availability, and a growing population. With less than 0.052 hectares of land per capita in Bangladesh and limited usable freshwater resources globally, the region is at a pivotal point for strategic intervention. Intensively farmed soils in South Asia are increasingly degraded, highlighting the urgency of restoring organic matter and adopting balanced fertilization practices that combine organic and inorganic sources. These methods improve soil structure, nutrient retention, and crop yield potential, while also supporting carbon sequestration and climate resilience.

In response to this complex challenge, water conservation techniques—like rainwater harvesting and efficient irrigation—must be scaled, especially given erratic rainfall patterns. Technical solutions alone aren't enough: mobilizing policy support, promoting farmer education, and expanding access to resources are vital. Regional initiatives in India and Bangladesh offer promising frameworks, but broader cooperation across South Asia is needed. Empowering communities, aligning research with field practices, and fostering inclusive governance will be critical to safeguard soil and water health and ensure sustainable.

5.10 Need to Increase Nitrogen Use Efficiency for Crop Production and Environmental Protection

Nitrogen remains a cornerstone nutrient in global agriculture, indispensable for crop growth and food production. Yet, its use efficiency—defined as the proportion of applied nitrogen absorbed and utilized by crops which rarely exceeds 20–40%. The remainder, known as unutilized reactive nitrogen (Nr), escapes into the environment, degrading soil health, polluting water bodies, contributing to greenhouse gas emissions, and exacerbating climate change. The challenge is particularly acute in South Asia, where intensive farming practices and generous fertilizer subsidies drive excessive nitrogen application, often with diminishing returns.

Some farm level proposed solutions are-split application of nitrogen, deep placement of urea, integrated nutrient management, use of nitrification and urease inhibitors, legume-based rotations, genetic crop improvement, and biochar amendments. Global research communities are now focused on developing location-specific best management practices to elevate NUE and cut nitrogen losses by 50%. A holistic strategy—one that aligns with local resources, farmer preferences, cost considerations, and ecological balance—will be key to transforming nitrogen management into a pillar of sustainable agriculture and environmental stewardship.

5.11 Case Studies

Gazipur Agriculture University completed a number stakeholders' engagement activities under the activities of the experiment and agronomic testing for finding N solutions. One of the well appreciated programs was the organization of an 'International Workshop on Reactive Nitrogen Management' on 22 March 2023. About 200 participants which include scientists from different research institutes, universities, agriculture extension, ministries, non-government organizations/civil societies, farmers attended the workshop. SANH partners from UK, India, Pakistan and Sri Lanka virtually connected and presented their research findings. News published in the national dailies are attached herewith.

https://www.dhakatribune.com/bangladesh/2023/03/22/use-of-nitrogen-fertil izer-rising-raises-concerns

DhakaTribune

Use of nitrogen fertilizer rising, raises concerns

Excessive and ineffective use of nitrogen fertilizers raises the cost of production and brings manifold threats to the environment, experts say

Workshop on reactive nitrogen management held at the Bangabandhu Sheikh Mujibur Rahman Agricultural University (BSMRAU) **Dhaka Tribune**.

Md. Raihanul Islam Akand, Gazipur. Published: March 22, 2023 11:12 PM.

Experts voiced concern over the excessive use of nitrogen fertilizer at an international workshop on reactive nitrogen management held at the Bangabandhu Sheikh Mujibur Rahman Agricultural University (BSMRAU) in Gazipur on Wednesday. The workshop was held as part of the research and stakeholders' engagement activities of the South Asian Nitrogen Hub (SANH). A wide range of stakeholders, farmers, researchers, academicians, postgraduate students, government officials, civil society members attended the workshop where scientists from Bangladesh, India, Pakistan, Sri Lanka and the UK presented their research activities. SANH is a research project of eight South Asian countries coordinated by the UK Centre for Ecology and Hydrology (CEH), funded by UK Research and Innovation, through the Global Challenge Research Fund (GCRF). SANH Bangladesh Lead Prof Dr. Md. Mizanur Rahman gave the welcome speech. Prof Dr. Mark Sutton, the hub director, virtually joined the event from the UK and delivered his speech on nitrogen challenges and solutions across the globe.

Habibun Nahar, deputy minister at the Ministry of Environment, Forest and Climate Change, was present at the event as the chief guest. She said excessive and ineffective use of nitrogen fertilizers raises the cost of production and brings manifold threats to the environment. She further emphasized the optimization of nitrogen fertilizer application for sustainable crop production and environmental protection. Special guest at the event, Treasurer of BSMRAU Tofayel Ahamed stated that unexploited reactive nitrogen degrades soil health, augments environmental menace, and increases greenhouse gas emission that contributes towards global warming and climate change. Prof Dr. Md. Giashuddin Miah, vice chancellor of BSMRAU, chaired the workshop.

5.11.1 Key Contributions of the Activity Conducted Under Experimental Testing and Agronomic Solutions for N Management

(1) Deep placement of urea super granule (USG) and biochar are identified as the best management options to increase rice yield, N use efficiency & farmer's income, with less N loss,

(2) Irrespective to rice seasons grain yield increased by 15–17% because of biochar application with N fertilizer and 8–9% from deep placement of USG when compared with recommended N rate as prilled urea,

(3) Compared to the recommended N, biochar with recommended N increased recovery N use efficiency (REN) 43–51%. Compared to the recommended N, the USG increased REN by 32–74%,

(4) Biochar with recommended N and USG had the greatest potential to reduce N losses without loss of rice yields,

(5) Leaching losses of nitrogen decreased by about 50% from biochar and deep placement of USG application compared with recommended rates of N application,

(6) Volatilization losses were also found lower at the biochar with N and USG application compared with recommended N application,

(7) In the Boro season, biochar and USG contributed to reduce N_2O–N emission by 24% and 56%, respectively compared to recommended N, while these values for T. Aman season were 31% and 47%, respectively.

(8) Findings Communicated with different stakeholders organizing a number workshops, seminars, conferences and raised awareness and created huge impact in the scientific, academic, students, policy makers and farmers community about efficient nitrogen management for sustainable agriculture and environment.

5.11.2 Village Level Engagement for N Solutions

Under the activity we worked closely with villages, conducted surveys to determine the main cause of N waste as well as possible challenges in adopting different practices. GAU also conducted experiments at the farmers' field using the best performed treatments at the on-station including farmers' practice. GAU selected Tokenagar village for conduction of survey and experimentation. Tokenagar is located in between 24°02' and 24°16' north latitudes and in between 90°30' and 90°42' east longitudes. Tokenagar is the Technology Village of GAU where research findings and technologies developed by the university are disseminated among the farmers of the village. The village is under the agro-ecological zone of Old Brahmaputra Floodplain. The soil is good for growing different crops. The farmers of the village are progressive and early adopters, with a literacy rate of about 44%. Traditionally, cropping, livestock and fisheries are the dominant agricultural activities practiced in this village. Rice based cropping systems dominate, and rice-rice, rice–wheat and rice-vegetable rotations are observed. Most of the farmers in this village are smallholders. Farmers apply inorganic fertilizers combined with different organic fertilizers. Locally available organic fertilizers include poultry manure, vermicompost and cow dung. Urea is applied in higher doses than other inorganic and organic fertilizers. Broadcasting of prilled urea is widely practiced by the farmers.

5.11.3 Conduction of Household Survey at Tokenagar

Nitrogen management for crop production at the village level is a great concern across the South Asian countries. Better understanding and adoption of best options of nitrogen management for sustainable agriculture is crucial. Therefore, for raising awareness about nitrogen several stakeholders' engagement activities, household surveys and research at the village level were conducted (Fig. 5.12). In Bangladesh, such activities were conducted at Tokenagar village in Kapasia of Gazipur district during 2021–2023. Before conduction of stakeholders' survey and experimentation at the farmers' field, an 'Ethical Guidelines and Safeguarding Policy for Research' have been prepared and approved by the Academic Council and Syndicate of GAU. Then, an initial survey (Survey 0) was conducted with the households of the village to stratify the nature of farming activities undertaken in the village. Then, a crop survey was conducted with different farmer's groups to know the NUE in the crop management processes at the village level.

The 1st study was started with an initial survey (Survey 0). This survey was conducted with the households of the village and farmers were classified into four clusters based on farm size (Fig. 5.13). Each of the Cluster 1 and 2 further divided into two subclusters. After that, seven households were randomly surveyed from each group in the crop survey, and the findings indicated that the farmers group in Cluster 1 subcluster 1 had the largest partial N budget. Conversely, Cluster 3 showed the least

Fig. 5.12 Conduction household survey at Tokenagar village, Gazipur, Bangladesh

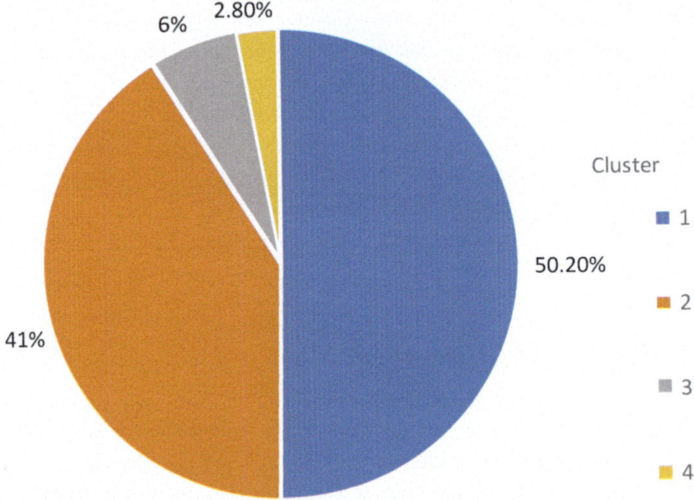

Fig. 5.13 Clustering of household based on farm sizes (1 = Marginal farmers (0.02–0.2 ha), 2 = Small farmers (0.2–1 ha), 3 = Medium farmers (1–3 ha) and 4 = Small farmer with commercial poultry farm)

amount of partial N budget. However, despite this, Cluster 3 farmers achieved the highest fertilizer and total input NUE. This success was attributed to their superior access to resources, modern technology, and infrastructure, allowing them to adopt more effective N management strategies. In contrast, Cluster 1 Subcluster 1 and Cluster 4 exhibited the lowest NUE levels due to resource limitation and lack of effective N management practices. So, NUE varied among different farmer groups, suggesting the possibility of improving NUE in the crop management processes at the village level (Fig. 5.14).

The partial N budget of different clusters is demonstrated (Table 5.2). This budget accounted for synthetic and organic N fertilizer as N inputs in crop production. Other sources of N input, like lighting, biological N fixation (BNF), and rain fall,

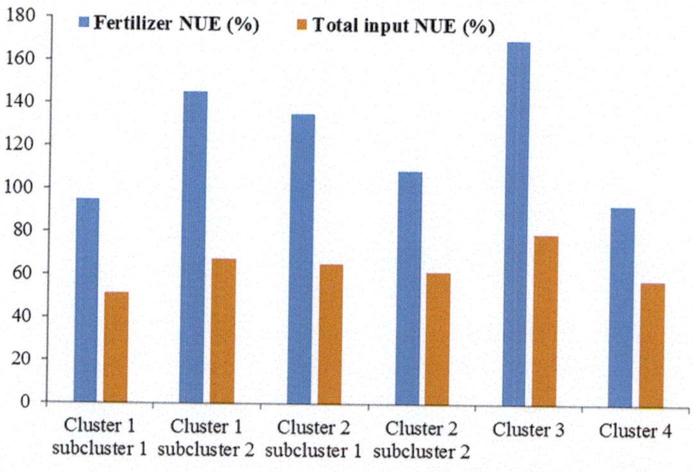

Fig. 5.14 Fertilizer nitrogen use efficiency and total input use efficiency in different farmer's groups

Table 5.2 Partial nitrogen budgeting for crops in different farmer's groups

Name of farmer groups	Synthetic N-fertilizer used (kg ha^{-1} yr^{-1})	Organic N-fertilizer used (kg ha^{-1} yr^{-1})	N uptake in harvested products (kg ha^{-1} yr^{-1})	N uptake in crop residues (kg ha^{-1} yr^{-1})	Partial N budget (kg ha^{-1} yr^{-1})
Cluster 1 subcluster 1	260	133ab	107ab	60ab	226a
Cluster 1 subcluster 2	141	158ab	115ab	65ab	118ab
Cluster 2 subcluster 1	166	183a	136a	77a	135ab
Cluster 2 subcluster 2	139	95b	92b	52b	91ab
Cluster 3	143	141ab	144a	82a	57b
Cluster 4	192	120ab	112ab	66ab	135ab
LSD (5%)	NS	74.61	42.16	24.45	140.43

were not accounted for as N input. N uptake in harvested products and crop residues was considered N output. Other sources of N output, like leaching and gaseous N losses, were not accounted for as N output. The highest amount of partial N budget was observed in Cluster 1 subcluster 1 (226 kg ha^{-1} yr^{-1}) which was statistically similar to the amount of Cluster 2 subcluster 1 (135 kg ha^{-1} yr^{-1}), Cluster 4 (135 kg ha^{-1} yr^{-1}), Cluster 1 subcluster 2 (118 kg ha^{-1} yr^{-1}) and Cluster 2 subcluster 2 (91 kg ha^{-1} yr^{-1}). The lowest amount of partial N budget was observed in Cluster 3 (57 kg ha^{-1} yr^{-1}).

Table 5.3 Effect of NUE in crop production on soil nitrogen

Si no	Interpretation	NUE (%) in crop production system
1	Soil N mining	>100
2	Risk of soil N mining	90–100
3	Balanced N fertilizer	70–90
4	Risk of N losses	50–70
5	High risk of N losses	<50

Both high and low NUE in the crop production system are harmful for the soil and the entire environment (Table 5.3). High NUE can cause soil N mining, while low NUE can cause N losses. It is observed that 70–90% NUE in crop production system confirms balanced N fertilization. So, the farmer's group of Cluster 3 enabled to implement efficient N management practices. On the other hand, Cluster 1 Subcluster 1 and Cluster 4 showed the lowest NUE. Farmers in Cluster 3 were categorized as a medium farmer group and they had 1–3 ha of land. So, they had access to resources, modern technology, and infrastructure, which could potentially improve their NUE. On the other hand, farmers of other clusters were often faced resource constraints, limited access to modern technologies, and financial limitations, which may have affected their ability to implement efficient nitrogen management practices. Overall, the data suggested that farmers in cluster 3 performed well in terms of both fertilizer NUE and total input NUE, demonstrating efficient N management practices. On the other hand, Cluster 1 Subcluster 1and Cluster 4 showed the lowest nitrogen use efficiency.

5.11.4 Village Level N Management at Tokenagar

An on-farm experiment was conducted in Tokenagar village, Gazipur, Bangladesh to evaluate the effectiveness of various nitrogen (N) management strategies on rice yield and nitrogen use efficiency (NUE) across two distinct growing seasons—Boro (dry season with full irrigation) and T. Aman (wet season with supplemental irrigation). The study also assessed the impact of these treatments on the physicochemical properties of rice field soils. Five nitrogen management treatments were implemented: a control (zero N), farmers' practice using broadcasted prilled urea (PU), recommended dose of nitrogen as PU based on soil testing (RDN), biochar (2 t ha^{-1}) combined with recommended PU (BRDN), and deep placement of urea super granules (USG) into the root zone. Grain yield and NUE indicators were recorded for both rice seasons.

Results revealed that the BRDN treatment substantially improved soil health—reducing bulk density, increasing water-holding capacity, and enhancing soil pH by 4.9%. The USG treatment also improved pH by 2.8%. Water-stable aggregates significantly increased in BRDN (148–157%) and USG (54–111%) treatments compared

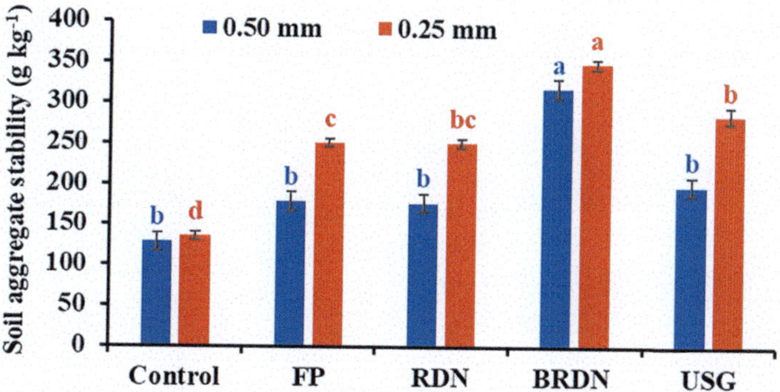

Fig. 5.15 Soil aggregate stability as affected by different N management practices (*Source* Al-Amin et al. 2024b). Different letters on the bar of respective parameter indicate significant difference

to the control, particularly in smaller particle sizes (0.25 mm) (Fig. 5.15). Biochar notably boosted soil carbon content by 18%, contributing to carbon sequestration goals. Elevated microbial biomass carbon and nitrogen further confirmed biochar's value for long-term soil sustainability (Fig. 5.16). A Field Day was observed as a Result Demonstration for the farmers of Tuknagar village (Fig. 5.17). Agronomically, USG and BRDN treatments improved grain yield by 34% and 30% during Boro and by 31% and 22% during T. Aman season, respectively, relative to farmer-practice (Table 5.4). These treatments also delivered marked improvements in agronomic efficiency and nitrogen recovery. Economically, USG deep placement yielded the highest net returns and benefit–cost ratio—3.2 times in Boro and 1.8 times in T. Aman. However, its adoption remains limited due to labor-intensive application methods and lack of efficient tools. While biochar entails higher upfront costs, its environmental benefits—such as soil carbon enrichment and reduced nitrogen losses—make it a valuable long-term investment. Policy support and affordable access to biochar and USG applicators are necessary to encourage adoption of these environmentally friendly technologies at the farm level.

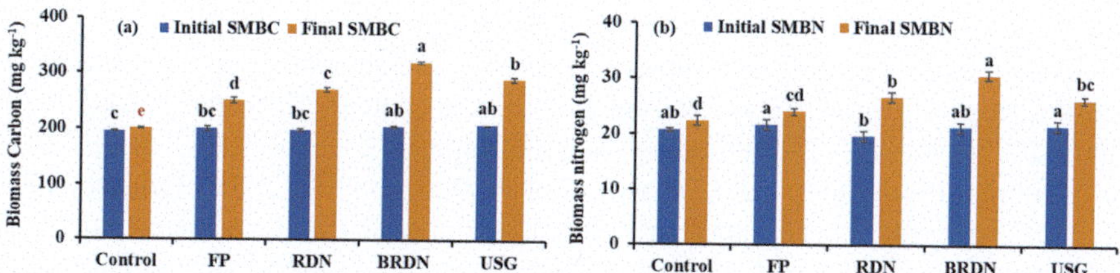

Fig. 5.16 Effects of different treatments on **a** soil microbial biomass carbon (SMBC) and **b** soil microbial biomass nitrogen (SMBN) of post-harvest soil; Vertical bar on the column indicates standard error (*Source* Al-Amin et al. 2024b). Different letters on the bar of respective parameter indicate significant difference

Fig. 5.17 Field-day observed at the Tokenagar village to show them the performance of N management practices and discussion with farmers about best management of cow dung for efficient N utilization

Table 5.4 Effect of different N management practices on grain, straw and biological yields of *Boro* and *T. Aman rice* at the farmer's field in 2023 (Al-Amin et al. 2025)

Treatments	Yield of *Boro rice* (t ha^{-1})			Yield of *T. Aman rice* (t ha^{-1})		
	Grain	Straw	Biological yield	Grain	Straw	Biological yield
Control	3.87e	4.63d	8.50d	2.83e	3.31d	6.14e
FP	4.69d	5.82c	10.51c	3.56d	4.24c	7.80d
RDN	5.50c	6.26b	11.76b	4.04c	4.67b	8.71c
BRDN	6.11b	6.37ab	12.48a	4.36b	4.75b	9.11b
USG	6.29a	6.48a	12.77a	4.65a	4.96a	9.61a
LSD (5%)	0.17	0.17	0.31	0.23	0.17	0.31
S.E. (±)	0.07	0.07	0.14	0.09	0.07	0.13

GAU completed a number stakeholders' engagement activities under the village level activity. We organized 'Nitro Champ Competition 2023' for GAU students as an awareness raising program on nitrogen at GAU on 16 August 2023 under the UKRI GCRF South Asian Nitrogen Hub. About 150 students and 20 faculty members attended the event. Photographs of the event are attached herewith (Fig. 5.18).

GAU organized a workshop entitled 'Awareness raising campaign on sustainable use of nitrogen fertilizer in agriculture and climate change effect mitigation' on 01 March 2023 at the Toke Narendra High School, Kapasia, Gazipur, Bangladesh. About 100 students, 15 teachers, and 15 others attended the event (Fig. 5.19). On December 22, 2021 GAU organized a workshop entitled 'Sustainable Fertilizer Management' as incentive for the farmers of Tokenagar village. About 100 farmers from the Tokenagar village attended the workshop (Fig. 5.20). GAU organized a stakeholders' dialogue as a host on 'Sustainable Nitrogen Management and Climate Change Effect Mitigation' at the Auditorium of Bangladesh Institute of Nuclear Agriculture (BINA), Cumilla on October 25, 2021. Farmers of Chandinaupazila, scientists of BARI, BRRI, BINA, SRDI, BARD, and extension personnel of DAE, academicians of GAU and SAU

Fig. 5.18 Nitro Champ Competition 2023' for GAU students as an awareness raising program on nitrogen at GAU held on 16 August 2023

were attended the dialogue. About 150 participants attended the dialogue. News of the Stakeholders' dialogue was published by the National Daily News Paper 'New Nation' is attached herewith.

Key Outputs of the Village Level Study on N Management

(1) Similar to on-station experiment at GAU the farmers' field trial also confirmed that the deep placement of urea super granule (USG) and biochar increased rice yield, N use efficiency and farmer's income.

(2) Compared with the farmer-practice, application of USG and biochar with recommended N increased grain yield by 34% and 30% in the *Boro*, and by 31% and 22% in the *T. Aman* seasons, respectively.

(3) Irrespective of rice season, USG and biochar resulted in a spectacular increase in agronomic and recovery efficiencies of applied N compared to farmer-practice or the recommended N as broadcasting of prilled urea.

(4) Urea deep placement appeared to be the most economically viable option, with a 3.2-fold increase in net return in Boro and 1.8-fold increase in *T. Aman* compared to farmer-practice.

(5) Increment of 18% carbon for the biochar treatment is notable in terms of carbon-negative economy

Fig. 5.19 Awareness raising campaign on sustainable use of N fertilizer in agriculture on 01 March 2023 at the Toke Narendra High School, Gazipur, Bangladesh

(6) Soil aggregates, microbial biomass carbon and biomass nitrogen increased significantly when biochar with recommended N and USG applied compared with the sole recommended N application.

(7) Findings Communicated with different stakeholders organizing a number workshops, seminars, conferences and raised awareness and created huge impact in the scientific, academic, students, policy makers and farmers community about efficient nitrogen management for sustainable agriculture and environment.

5.11.5 Village Level N Management in Mumtajpur, Haryana, India

A village level study was carried out in an Indian village under the South Asian Nitrogen Hub project. At village level, all the information related to farmers, crops and livestock was collected using the various survey formats digitally. Two surveys 1a and 1b were conducted to collect the information of crops and livestock respectively, at Mumtajpur village cluster in Patuaudi block, Gurugram District, Haryana, India (Fig. 5.21). In survey 1a, 106 households were selected based on different criteria such as landholdings, area of cultivable land, crops, types of agriculture, irrigation

Fig. 5.20 Workshop on sustainable fertilizer management at GAU

facilities, education level and others out of 238 households. Out of 106 households, 36 households were selected from 5 different categories for the further detailed survey 1b on crop growth and livestock management. The survey work was carried out digitally using the ArcGIS Survey123 app.

Fate of Nitrogen Sources at Village Level

The fate of nitrogen sources at the village level depends on various factors, such as the type and amount of nitrogen sources available, the management practices of farmers, and the soil and environmental conditions. Nitrogen sources commonly used in agriculture include synthetic fertilizers, animal manure, crop residues, and nitrogen-fixing crops. Synthetic fertilizers typically contain high concentrations of nitrogen in the form of ammonium or nitrate (Fig. 5.22). When applied to the soil, these forms of nitrogen can be taken up by plants, leached into groundwater, drainage in surface water bodies, or lost to the atmosphere through volatilization. Overuse of synthetic fertilizers can lead to soil acidification, nutrient imbalances, greenhouse gas emissions, eutrophication, and environmental pollution. Animal manure is a common nitrogen source in rural areas. Manure can provide valuable nutrients to crops and improve soil health when adequately managed. However, manure can contaminate water sources and contribute to greenhouse gas emissions if not handled properly.

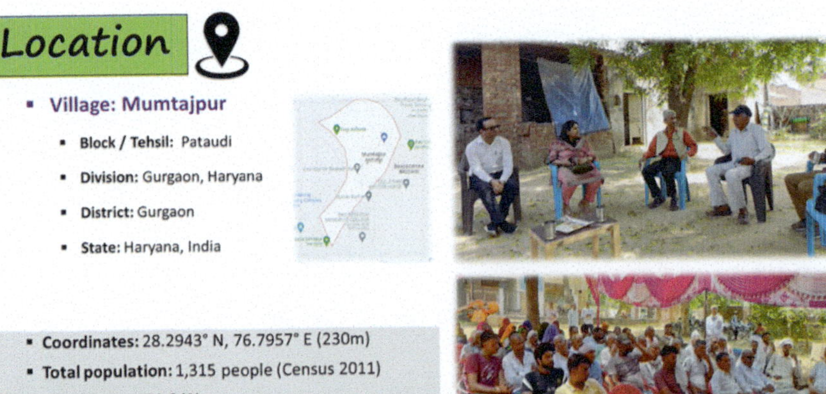

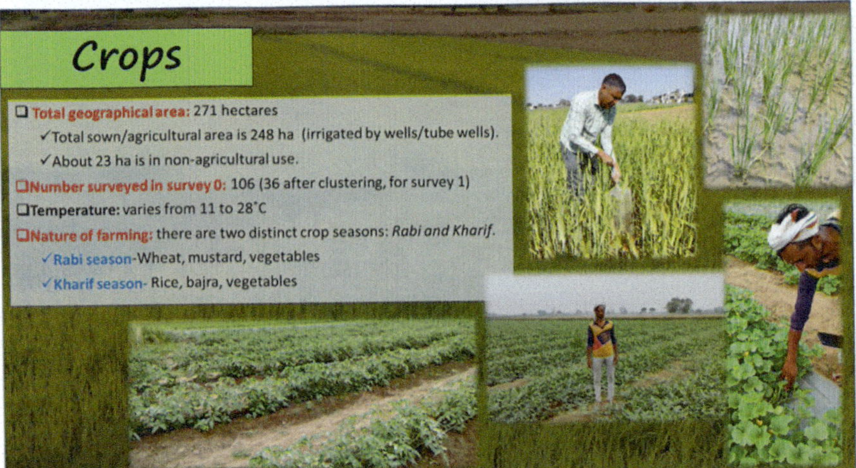

Fig. 5.21 Village level N management in Mumtajpur, Haryana, India

Crop residues, such as straw or corn stalks, can also provide a source of nitrogen for crops. When left on the soil surface, these residues can improve soil structure and increase organic matter. However, nitrogen can be immobilized and unavailable to crops if incorporated slowly into the soil. Proper management of nitrogen-fixing crops can help reduce the need for synthetic fertilizers and improve soil health. Overall, the fate of nitrogen sources at the village level depends on a complex interaction of factors. Proper management practices, such as timing and application rate, can help maximize the benefits of nitrogen sources while minimizing negative impacts on the environment.

Fig. 5.21 (continued)

Fig. 5.22 Sources of reactive N losses in villages

Farmers Training and Stakeholder Engagement to Reduce Reactive N Losses

The advantages of inexpensive and simple N management tools such as Leaf Color Chart (LCC) based application N fertilizer application in crops, foliar application of nanourea fertilizer to crops, benefits of biofertilizers application in rice and other crops, managing insect-pest attack using pheromone traps, identifying common diseases, vermicomposting, green manuring and composting are required for successful management of resources especially Nitrogen, thereby reducing reactive N losses.

5.12 Overall Key Outputs

Stakeholders involved in sustainable nitrogen management initiatives have achieved impactful outcomes through collaborative efforts. A policy document for nitrogen uses and accompanying ethical research guidelines were developed to guide future practices and ensure responsible implementation. Technological innovation led to the creation of an Android-based mobile application designed to enhance nitrogen use efficiency (NUE) among farmers. Field research revealed that traditional nitrogen management methods resulted in significant nitrogen losses, whereas biochar application and deep placement of urea super granules (USG), and LCC based N fertilizer application proved to be the most effective techniques for boosting rice yield, NUE, and farmer income while minimizing environmental harm (Fig. 5.23). Despite biochar's high cost, deep placement of USG yielded the highest economic returns for farmers. Adoption of these environment-friendly methods remains low due to limited farmer awareness, which is now being addressed through large-scale training programs involving over 500 farmers, 100 extension workers, 500 students, and 100 additional stakeholders. The inclusion of biochar as a long-term soil amendment also opens the door for carbon-negative agricultural practices. Finally, the initiative strengthened international collaboration, forging strong ties between scientists from South Asia and the UK, which lays the foundation for solving broader agricultural and environmental challenges through shared innovation and knowledge exchange (Fig. 5.24).

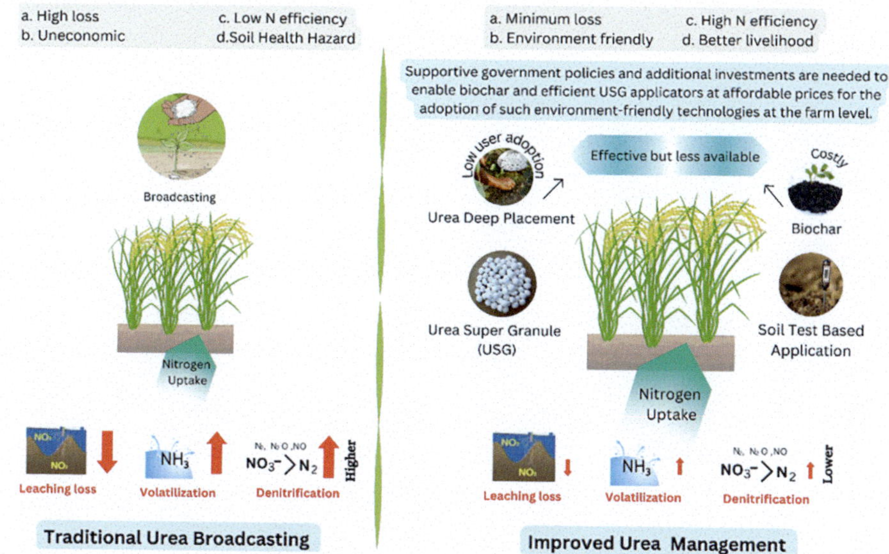

Fig. 5.23 Shifting from traditional N management to improved N management

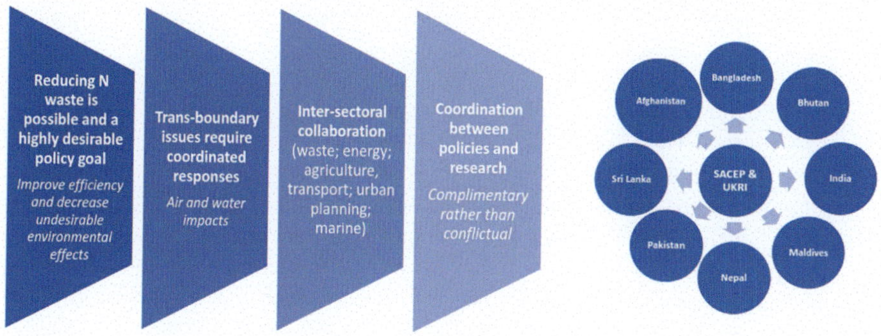

Fig. 5.24 Coordinated efforts of South Asia and UK might catalyse other regions of the world to respond and act in addressing N challenges

Chapter 6
Key Challenges and Way Forward

Nitrogen air pollution mainly from nitrogen oxides (NOx) and ammonia (NH_3) is a matter of concern due to its adverse effects on human health and environment (Sutton et al. 2017). NO_2 is primarily emitted from fossil fuel combustion and NH_3 largely from agricultural activities. NOx also contributes to greenhouse effect through tropospheric ozone formation, and is linked to severe respiratory diseases, while NH_3 plays a major role in $PM_{2.5}$ pollution, causing chronic health issues. Global NO_2 levels have risen in different regions between 2005 and 2018 especially in India, China, and the Middle East. India's urea consumption has surged since the Green Revolution, intensifying RNS emissions and linking nitrogen pollution to climate change through CO_2 and N_2O emissions. Recognizing these challenges, the UN Environment Assembly in 2021 adopted a resolution for sustainable nitrogen management, urging global reductions in nitrogen waste by 2030.

6.1 Nitrogen Pollution Mitigation

Recent studies highlight rising ambient RNS levels, underscoring the urgent need for targeted monitoring and policy interventions on fertilizers and fuels. Some of the strategic actions are suggested below to address the RNS issues in South Asia-

1. **Develop a Comprehensive Assessment Framework**

    - Create an integrated plan to assess reactive nitrogen species (RNS) emissions and deposition in terms of urbanization and industrialization progress.RNS are a group of highly reactive molecules derived from nitrogen that play crucial roles in both biological systems and environment e.g., NOx and NH_3. This plan must include a design followed by the South Asian Nitrogen Hub project covering agriculture, air water and soil pollution, energy, biodiversity, climate change and policy integration aspects.

U. C. Kulshrestha et al., *Atmosphere-Biosphere Interactions of Reactive Nitrogen in South Asia*, Springer Briefs in Interdisciplinary Geosciences – Asia-Pacific,
https://doi.org/10.1007/978-3-032-17194-8_6

- Include aspects such as emissions, abundance, transport, transformation, scavenging, environmental impacts, and forest nitrogen pools.

2. **Enhance Data Quality and Protocols**

- Establish standardized protocols for RNS emissions and deposition data collection.
- Promote inter-laboratory comparison exercises for high quality measurements.
- Initiate development of reference standards for consistent measurements and distribute among different participating groups.
- Develop metadata directories for different parameters and regions.

3. **Establish a Long-Term Monitoring Network**

- Set up a network of diverse monitoring sites across countries to study nitrogenous gases, aerosols, atmospheric deposition, water pollution, soil pollution and ecological changes etc.
- Ensure the network supports long-term data collection for policy-relevant research. Long-term data is the backbone of smart, resilient policy formulation especially in air pollution, climate change, and public health, where trends are unfolded over years or decades.
- Such network will help in the assessment of RNS using own long-term data.

4. **Link Deposition Studies with Trajectory Analysis**

- Integrate atmospheric trajectory modelling with atmospheric deposition to understand local, trans-boundary, and long-range transport of RNS.
- Launch a regional pollution monitoring initiative under SAARC to track import/export of pollutants such as SO_2, NOx and submicron aerosols across the region.

5. **Prioritize Impact Assessments**
Conduct focused studies on the environmental and health impacts of various RNS compounds. RNS such as NH_3 and NO_x have significant health impacts, especially in densely populated regions like South Asia where emissions from agriculture, vehicles, and industry are high. These reactive nitrogen species contribute to both direct toxicity and secondary pollutant formation, notably $PM_{2.5}$ and ozone, which are linked to serious health problems.
6. **Quantify Indoor Dry Deposition Budgets**
Measure concentrations and dry deposition of gases and aerosols, especially ammonia (NH_3) and ammonium (NH_4^+), in indoor environments. Sometimes indoor air has very high NH_3. High concentrations of NH_3 are responsible for eye, nose, and throat irritation. NH_3 also gives respiratory distress and mucous membrane damage.
7. **Form a Specialized Task Force**

- Create a dedicated team to investigate gas-aerosol interactions, scavenging, transport, evapotranspiration, deposition, and uptake processes.

- Emphasize dry deposition studies to reduce uncertainties in regional Nr budgets.

8. **Build a Collaborative Modeling Consortium**

   - Assemble a multidisciplinary modeling group of active scientists and research teams.
   - Include socio-economic experts to help translate scientific findings into actionable policy recommendations. Trans-disciplinary assessment of RNS may play an important role in providing a realistic picture of nitrogen distribution across regions.

6.2 N Management Future Priorities

Nitrogen fertilizers are essential for global food security, but their inefficient use leads to following key challenges for us:

- Overuse and blanket fertilizer recommendations.
- Low nitrogen use efficiency (30–40%) in many cropping systems across the world.
- Environmental pollution (nitrate leaching, ammonia volatilization, nitrous oxide emissions).
- Soil health degradation and rising production costs for farmers.
- Low adoption of efficient technologies (deep placement, inhibitors, coated urea).
- Lack of farmer awareness and weak policy support.
- Climate change impacts intensify N losses.

To maintain harmony between crop production and environmental protection sustainable N management strategies are required as outlined below:

1. **Adoption of Site-Specific Nutrient Management (SSNM)**

   - Application of N as per crop demand, soil fertility status, and climatic conditions.
   - Practice soil testing and decision-support tools to optimize fertilizer rates.

2. **4R Nutrient Stewardship (Right Source, Right Rate, Right Time, Right Place)**

   - Provide the correct fertilizer type (urea, coated urea, organic amendments).
   - Use appropriate doses instead of blanket recommendations.
   - Synchronize split application of N with crop growth stages (tillering, panicle initiation, etc.).
   - Ensure right placement methods (e.g., deep placement, banding) to minimize losses.

3. **Use of Enhanced Efficiency Fertilizers**

- Promote urease and nitrification inhibitors, biochar, slow-release fertilizers, and polymer/sulfur-coated urea to reduce volatilization and leaching.
- Support farmer access through subsidies or incentives for eco-friendly fertilizers.

4. **Integration of Organic and Inorganic Sources**
   - Combine inorganic fertilizers with organic nutrient sources and amendments like compost, green manures, crop residues, biochar and biofertilizers to improve soil health and nutrient cycling.
   - Supply soil organic matter which contributes to augment microbial activity, and nutrient use efficiency.

5. **Precision Agriculture Technologies**
   Use of drones, sensors, and GIS-based tools for real-time monitoring of crop N status. Drones equipped with multispectral and hyperspectral sensors can detect canopy reflectance, chlorophyll levels, and normalized difference *vegetation* index (NDVI), enabling site-specific N recommendations. Recently, real-time monitoring tools are increasingly affordable and scalable, even for smallholder farmers.
6. **Water and Irrigation Management**
   - Synchronize irrigation scheduling with N application to reduce leaching and denitrification.
   - Adopt alternate wetting and drying (AWD) in rice fields to reduce water requirement, minimize N losses and greenhouse gas emissions.

7. **Awareness and Capacity Building**
   - Train farmers on improved fertilizer management practices.
   - Promote demonstration plots and farmer field schools to encourage adoption of new technologies.
   - Consumer awareness is equally important for letting them the nitrogen losses (footprints) associated with different foods in their diets. This awareness would open a new avenue for N management.

8. **Policy Support and Incentives**
   - Provide subsidies for eco-friendly fertilizers and precision farming tools.
   - Introduce regulatory frameworks to discourage excessive and inefficient N use.
   - Encourage carbon credit and payment for ecosystem services for farmers who reduce N losses.

9. **Research and Innovation**
   - Develop climate-smart N management technologies suitable for smallholder farmers.
   - Breed crop varieties with higher nitrogen use efficiency (NUE).
   - Promote integrated soil fertility management (ISFM) approaches.

10. **Monitoring and Environmental Protection**
    - Establish monitoring systems for nitrate leaching, ammonia volatilization, and nitrous oxide emissions.
    - Strengthen policies for water quality protection and reduction of air pollution from N fertilizers.

In summary, sustainable N-fertilizer management requires a holistic approach that combines technological innovations (precision tools, enhanced efficiency fertilizers), ecological practices (organic matter recycling, integrated management), farmer education, and supportive policies. By moving toward site-specific and efficient use of N, agriculture can confirm higher crop yields, protect the environment and contribute to long-term food security. More about nitrogen management is available in the guidance documents recently released by the International Nitrogen Management System (INMS) for policymakers, practitioners, and stakeholders. These resources aim to support effective nitrogen management through evidence-based approaches and practical implementation strategies (https://www.unep.org/events/webinar/new-international-guidance-and-tools-sustainable-nitrogen-management).

References

Abalos D, Jeffery S, Sanz-Cobena A, Guardia G, Vallejo A (2014) Meta-analysis of the effect of urease and nitrification inhibitors on crop productivity and nitrogen use efficiency. Agr Ecosyst Environ 189:136–144

Abrol YP, Adhya TK, Aneja VP, Raghuram N, Pathak H, Kulshrestha U, Sharma C, Singh B (eds) (2017) The Indian nitrogen assessment. Elsevier, USA. ISBN: 9780128118368

Ajibola BO, Fatoki P (2017) Adoption of urea deep placement recommended practices among rice farmers in Niger State

Akiyama H, Yan X, Yagi K (2010) Evaluation of effectiveness of enhanced-efficiency fertilizers as mitigation options for N_2O and NO emissions from agricultural soils: meta-analysis. Glob Change Biol 16(6):1837–1846

Alam MS, Khanam M, Rahman MM (2023) Environment friendly nitrogen management practices in wetland paddy cultivation. Front Sustain Food Syst 7:94. https://doi.org/10.3389/fsufs.2023.1020570

Alam MS, Xia W, Jia Z (2014) Methane and ammonia oxidations interact in paddy soils. Int J Agric Biol 16:365–370

Al-Amin HM, Rahman MM, Alam MSF, Smith J, Islam M, Ress RM, Adhya TK, Sutton MA, Bealey WJ, Miah MG, Islam MR (2025) Village level fertilizer management for increasing nitrogen use efficiency, rice yield and household income. Arch Agron Soil Sci 71(1):1–19. https://doi.org/10.1080/03650340.2025.2457764

Al-Amin HM, Rahman MM, Meena RS, Biswas JC, Alam MS, Kamal MZU (2024a) Current scenario and challenges for agricultural sustainability. In: Rahman MM, Biswas JC, Meena RS (eds) Climate change and soil-water-plant nexus. Springer, Singapore. https://doi.org/10.1007/978-981-97-6635-2_14

Al-Amin HM, Rahman MM, Alam MS, Smith J, Sutton MA, Miah MG, Islam MR (2024b) Biochar addition coupled with nitrogen fertilization improves soil fertility, nitrogen use efficiency and rice yield. Bangladesh J Soil Sci 40(1):33–44

Albertsson I, Sjöholm J, Ter Beek J, Watmough NJ, Widengren J, Ädelroth P (2019) Functional interactions between nitrite reductase and nitric oxide reductase from Paracoccusde nitrificans. Sci Rep 9:17234

Anik AR, Eory V, Begho T, Rahman MM (2023) Determinants of nitrogen use efficiency and gaseous emissions assessed from farm survey: a case of wheat in Bangladesh. Agric Syst 206:103617. https://doi.org/10.1016/j.agsy.2023.103617

Anglade J, Billen G, Garnier J (2015) Relationships for estimating N_2 fixation in legumes: incidence for N balance of legume-based cropping systems in Europe. Ecosphere 6(3):1–24

U. C. Kulshrestha et al., *Atmosphere-Biosphere Interactions of Reactive Nitrogen in South Asia*, Springer Briefs in Interdisciplinary Geosciences – Asia-Pacific,
https://doi.org/10.1007/978-3-032-17194-8

Ayiti OE, Babalola OO (2022) Factors influencing soil nitrification process and the effect on environment and health. Front Sustain Food Syst 6:821994. https://doi.org/10.3389/fsufs.2022.821994

Azeem B, KuShaari K, Man ZB, Basit A, Thanh TH (2014) Review on materials and methods to produce controlled release coated urea fertilizer. J Control Release 181:11–21

Ban S, Matsuda K, Ohizumi T (2016) Wet deposition of nitrogen and sulfur at remote monitoring sites in Japan: long-term trends and regional characteristics. Atmos Environ 146:70–78. https://doi.org/10.1016/j.atmosenv.2016.06.006

BARI (2018) Development and dissemination of fertilizer deep placement applicator for increasing fertilizer use efficiency and farm productivity. PCR submitted to NATP, Phase 2, BARC, Dhaka 1215. Bangladesh Agricultural Research Institute, Gazipur 1701, Bangladesh

Beachley GM, Fenn ME, Du E, de Vries W, Bauters M, Bell MD, Kulshrestha UC, Schmitz A, Walker JT (2023) Monitoring nitrogen deposition in global forests. In: Du E, de Vries (eds) Atmospheric nitrogen deposition to global forests. Academic Press, pp 17–38. ISBN: 978-0-323-91140-5

Behera SN, Sharma M (2010) Investigating the potential role of ammonia in ion chemistry of fine particulate matter formation for an urban environment. Sci Total Environ 408:3569–3575

Biswas K et al (2008) Gaseous and aerosol pollutants during fog and clear episodes in South Asian urban atmosphere. Atmos Environ 42:7775–7785

Bhatia A, Cowan NJ, Drewer J, Tomer R, Kumar V, Sharma S, Pathak H et al (2023) The impact of different fertiliser management options and cultivars on nitrogen use efficiency and yield for rice cropping in the Indo-Gangetic Plain: two seasons of methane, nitrous oxide and ammonia emissions. Agr Ecosyst Environ 355:108593. https://doi.org/10.1016/j.agee.2023.108593

Bhatia A, Pathak H, Aggarwal PK, Jain N (2010a) Trade-off between productivity enhancement and global warming potential of rice and wheat in India. Nutr Cycl Agroecosyst 86:413–424. https://doi.org/10.1007/s10705-009-9304-5

Bhatia A, Sasmal S, Jain N, Pathak H, Kumar R, Singh A (2010b) Mitigating nitrous oxide emission from soil under conventional and no-tillage in wheat using nitrification inhibitors. Agric Ecosyst Environ 136(3–4):247–253

Bhatia A, Tomer R, Kumar V, Singh SD, Pathak H (2012a) Impact of tropospheric ozone on crop growth and productivity–a review. J Sci Ind Res 71(2):97–112

Bhatia A, Pathak H, Jain N, Singh PK, Tomer R (2012b) Greenhouse gas mitigation in rice–wheat system with leaf color chart-based urea application. Environ Monit Assess 184(5):3095–3107. https://doi.org/10.1007/s10661-011-2174-8

Bhatia A, Jain N, Pathak H (2013a) Methane and nitrous oxide emissions from Indian rice paddies, agricultural soils and crop residue burning. Greenhouse Gases Sci Technol 3(3):196–211

Bhatia A, Kumar A, Das TK, Singh J, Jain N, Pathak H (2013b) Methane and nitrous oxide emissions from soils under direct seeded rice. Int J Agric Stat Sci 9(2):729–736

Bhattacharyya R, Bhatia A, Das TK, Lata S, Kumar A, Tomer R, Singh G, Kumar S, Biswas AK (2018) Aggregate-associated N and global warming potential of conservation agriculture-based cropping of maize-wheat system in the north-western Indo-Gangetic Plains. Soil Tillage Res 182:66–77

Bhatt B, Chandra R, Ram S, Pareek N (2016) Long-term effects of fertilization and manuring on productivity and soil biological properties under rice (*Oryza sativa*)–wheat (*Triticum aestivum*) sequence in Mollisols. Arch Agron Soil Sci 62(8):1109–1122. https://doi.org/10.1080/03650340.2015.1125471

Billah MM, Mina KK, Sharif DA, Abdullah HM, Rahman MM (2024) Advances in the use of remote sensing techniques to assess crop nitrogen status. In: Rahman MM, Biswas JC, Meena RS (eds) Climate change and soil-water-plant nexus. Springer, Singapore. https://doi.org/10.1007/978-981-97-6635-2_4

Braker G, Conrad R (2011) Diversity, structure, and size of N_2O-producing microbial communities in soils—what matters for their functioning? Adv Appl Microbiol 75:33–70

Breuer L, Kiese R, Butterbach-Bahl K (2002) Temperature and moisture effects on nitrification rates in tropical rain-forest soils. Soil Sci Soc Am J 66(3):834–844

Chen X, Peltier E, Sturm BS, Young CB (2013) Nitrogen removal and nitrifying and denitrifying bacteria quantification in a stormwater bioretention system. Water Res 47(4):1691–1700

Connell JA, Hancock DW, Durham RG, Cabrera ML, Harris GH (2011) Comparison of enhanced-efficiency nitrogen fertilizers for reducing ammonia loss and improving bermudagrass forage production. Crop Sci 51(5):2237–2248

Coskun D, Britto DT, Shi W, Kronzucker HJ (2017) Nitrogen transformations in modern agriculture and the role of biological nitrification inhibition. Nat Plants 3(6):17074. https://doi.org/10.1038/nplants.2017.74

Cowan N, Bhatia A, Drewer J, Jain N, Singh R, Tomer R, Kumar V, Kumar O, Prasanna R, Ramakrishnan B, Kumar D, Bandyopadhyay SK, Sutton M, Pathak H (2021) Experimental comparison of continuous and intermittent flooding of rice in relation to methane, nitrous oxide and ammonia emissions and the implications for nitrogen use efficiency and yield. Agr Ecosyst Environ 319:107571. https://doi.org/10.1016/j.agee.2021.107571

Chakrabarti B, Bhatia A, Sharma S, Tomer R, Sharma A, Paul A, Kumar V, Sutton MA (2024) Nitrification and urease inhibitors reduce gaseous N losses and improve nitrogen use efficiency in wheat exposed to elevated CO_2 and temperature. Front Sustain Food Syst 8:1460994

CPCB (2009) https://cpcb.nic.in/openpdffile.php?id=VGVuZGVyRmlsZXMvVGVuZGVyXzE1Ml9UZW5kZXJfMTUyX0NBQVFNUzAxMC0wMS0xMi5wZGY=. Accessed 24 July 2025

Dakora FD, Aboyinga RA, Mahama Y, Apaseku J (1987) Assessment of N_2 fixation in groundnut (*Arachis hypogaea* L.) and cowpea (*Vigna unguiculata* L. Walp) and their relative N contribution to a succeeding maize crop in Northern Ghana. MIRCEN J Appl Microbiol Biotechnol 3(4):389–399

Datta A, Adhya TK (2014) Effects of organic nitrification inhibitors on methane and nitrous oxide emission from tropical rice paddy. Atmos Environ 92:533–545. https://doi.org/10.1016/j.atmosenv.2014.04.009

Dave AM, Mehta MH, Aminabhavi TM, Kulkarni AR, Soppimath KS (1999) A review on controlled release of nitrogen fertilizers through polymeric membrane devices. Polym-Plastics Technol Eng 38:675–711

Deng M, Hou M, Ohkama-Ohtsu N, Yokoyama T, Tanaka H, Nakajima K, Omata R, Bellingrath-Kimura SD (2017) Nitrous oxide emission from organic fertilizer and controlled release fertilizer in tea fields. Agriculture 7(3):29

Dentener F, Drevet J, Lamarque JF, Bey I, Eickhout B, Fiore AM, Hauglustaine D, Horowitz LW, Krol M, Kulshrestha UC, Lawrence M, Galy-Lacaux C, Rast S, Shindell D, Stevenson D, van Noije T, Atherton C, Bell N, Butler T, Cofala J, Collins B, Doherty R, Ellingsen K, Galloway J, Gauss M, Monranaro V, Müller JF, Pitari G, Rodriguez J, Sanderson M, Solmon F, Strahan S, Schultz M, Sudo K, Szopa S, Wild O (2006) Nitrogen and sulphur deposition on regional and global scales: a multimodel evaluation. Glob Biogeochem Cycles 20:GB4003. https://doi.org/10.1029/2005GB002672

Dieleman C, Bhatia A, Ravikumar A, Llovell F, Svane S, Tibrewal K, Zaelke D, Murphy A (2022) Opportunities beyond CO_2 for climate mitigation. One Earth 5(12):1308–1311. https://doi.org/10.1016/j.oneear.2022.11.017

Ding WX, Yu HY, Cai ZC (2011) Impact of urease and nitrification inhibitors on nitrous oxide emissions from fluvo-aquic soil in the North China Plain. Biol Fertil Soils 47(1):91–99

FAOSTAT (2023) Food and agriculture organization of the United Nations Statistics Division. http://faostat3.fao.org/home/E

FAOSTAT (2024) Food and agriculture organization of the United Nations Statistics Division. http://faostat3.fao.org/home/E

FAOSTAT (2025) Food and agriculture organization of the United Nations Statistics Division. https://www.fao.org/faostat/en/?#data

Fagodiya RK, Pathak H, Bhatia A, Jain N, Kumar A, Malyan SK (2020) Global warming impacts of nitrogen use in agriculture: an assessment for India since 1960. Carbon Manage 11(3):291–301

Folina A, Tataridas A, Mavroeidis A, Kousta A, Katsenios N, Efthimiadou A, Travlos IS, Roussis I, Darawsheh MK, Papastylianou P, Kakabouki I (2021) Evaluation of various nitrogen indices in N-fertilizers with inhibitors in field crops: a review. Agronomy 11:418. https://doi.org/10.3390/agronomy11030418

Galloway JN, Townsend AR, Erisman JW, Bekunda M, Cai Z, Freney JR, Sutton MA et al (2008) Transformation of the nitrogen cycle: recent trends, questions, and potential solutions. Science 320(5878):889–892

Gilsanz C, Báez D, Misselbrook TH, Dhanoa MS, Cárdenas LM (2016) Development of emission factors and efficiency of two nitrification inhibitors, DCD and DMPP. Agr Ecosyst Environ 216:1–8

Goos RJ (1985) Identification of ammonium thiosulfate as a nitrification and urease inhibitor. Soil Sci Soc Am J 49(1):244–248

Gould WD, Hagedorn C, McCready RCL (1986) Urea transformations and fertilizer efficiency in soil. Adv Agron 40:209–238

Gupta DK, Bhatia A, Das TK, Singh P, Kumar A, Jain N, Pathak H (2016) Economic analysis of different greenhouse gas mitigation technologies in rice–wheat cropping system of the Indo-Gangetic plains. Curr Sci:867–874

Gregory DI, Haefele SM, Buresh RJ, Singh U (2010) Fertilizer use, markets, and management. In: Pandey S et al (eds) Rice in the global economy: strategic research and policy issues for food security. International Rice Research Institute, pp 231–263

Halvorson AD, Snyder CS, Blaylock AD, Del Grosso SJ (2014) Enhanced-efficiency nitrogen fertilizers: potential role in nitrous oxide emission mitigation. Agron J 106(2):715–722

Hasan MK, Shahriar A, Jim KU (2019a) Water pollution in Bangladesh and its impact on public health. Heliyon 5(8). https://doi.org/10.1016/j.heliyon.2019.e02145

Hasan MA, Muraduzzaman M, Islam I, Miah GU, Rahman MM, Rahman A, Ahmed N, Ahmed Z (2019b) Spatiotemporal dynamics of new land development in Bangladesh coast and its potential uses. Remote Sens Appl Soc Environ 14:191–199. https://doi.org/10.1016/j.rsase.2019.04.001

Hasnat M, Alam MA, Khanam M, Binte BI, Kabir MH, Alam MS, Kamal MZ, Rahman GKM, Haque MM, Rahman MM (2022) Effect of nitrogen fertilizer and biochar on organic matter mineralization and carbon accretion in soil. Sustainability 14:3684. https://doi.org/10.3390/su14063684

Hassan MU, Aamer M, Mahmood A, Awan MI, Barbanti L, Seleiman MF, Bakhsh G, Alkharabsheh HM, Babür E, Shao J, Rasheed A, Guo-Qin H (2022) Management strategies to mitigate N_2O emissions in agriculture. Life 12(3):439. https://doi.org/10.3390/life12030439

He T, Yuan J, Xiang J, Lin Y, Luo J, Lindsey S, Liao X, Liu D, Ding W (2022) Combined biochar and double inhibitor application offsets NH_3 and N2O emissions and mitigates N leaching in paddy fields. Environ Pollut 292:118344

Heffer P, Gruère A, Roberts T (2013) Assessment of fertilizer use by crop at the global level. International Fertilizer Industry Association, Paris. https://www.fertilizer.org/wp-content/uploads/2023/01/AgCom.13.39-FUBC-assessment-2010.pdf

Huda A, Gaihre YK, Islam MR, Singh U, Islam MR, Sanabria J, Satter MA, Afroz H, Halder A, Jahiruddin M (2016) Floodwater ammonium, nitrogen use efficiency and rice yields with fertilizer deep placement and alternate wetting and drying under tipple rice cropping systems. Nutr Cycl Agroecosyst. https://doi.org/10.1007/s10705-015-9758-6

IFADATA (2020) Fertilizer N consumption pattern since 1961 in different regions of the world. Data source: http://ifadata.fertilizer.org/ucSearch.aspx

IFDC, International Fertilizer Development Center (2017) Rapid introduction and market development for urea deep placement technology for lowland transplanted rice: a reference guide. Muscle Shoals, Alabama

IPCC (2006) IPCC guidelines for national greenhouse gas inventories, Volume 4: Agriculture, Forestry and Other Land Use

IPCC (2014) Climate change 2014: synthesis report. In: Core Writing Team, Pachauri RK, Meyer LA (eds) Contribution of working groups I, II and III to the fifth assessment report of the intergovernmental panel on climate change. IPCC, Geneva, Switzerland, p 151

IPCC (2021) Climate change 2021: The physical science basis. Contribution of Working Group I to the Sixth Assessment Report of the Intergovernmental Panel on Climate Change [Masson-Delmotte, V., et al. (eds.)]. Cambridge University Press, Cambridge, UK and New York, NY, USA, 2391 pp

Islam M, Rahman MM, Alam MS, Rees RM, Rahman GM, Miah MG, Drewer J, Bhatia A, Sutton MA (2024a) Leaching and volatilization of nitrogen in paddy rice under different nitrogen management. Nutr Cycl Agroecosys 129:113–131

Islam M, Rahman MM, Meena RS (2024b) Consumption of biologically fixed green nitrogen and agricultural sustainability. In: Rahman MM, Biswas JC, Meena RS (eds) Climate change and soil-water-plant nexus. Springer, Singapore. https://doi.org/10.1007/978-981-97-6635-2_17

Islam SMJ, Mannan MA, Khaliq QA, Rahman MM (2018) Growth and yield response of maize to rice husk biochar. Aust J Crop Sci 12(12):1813–1819

Jiao X, Lyu Y, Wu X, Li H, Cheng L, Zhang C, Yuan L, Jiang R, Jiang B, Rengel Z, Zhang F, Davies WJ, Shen J (2016) Grain production versus resource and environmental costs: towards increasing sustainability of nutrient use in China. J Exp Bot 67:4935–4949. https://doi.org/10.1093/jxb/erw282

Katoch A, Kulshrestha UC (2021) Gaseous and particulate reactive nitrogen species in the indoor air of selected households in New Delhi. Environ Monit Assess 193(4):231. https://doi.org/10.1007/s10661-021-08991-6

Khalil M, Rosenani A, Van Cleemput O, Boeckx P, Shamshuddin J, Fauziah C (2002) Nitrous oxide production from an ultisol of the humid tropics treated with different nitrogen sources and moisture regimes. Biol Fertil Soils 36(1):59–65

Kim DG, Saggar S, Roudier P (2012) The effect of nitrification inhibitors on soil ammonia emissions in nitrogen managed soils: a meta-analysis. Nutr Cycl Agroecosyst 93(1):51–64

Kulshrestha U. 2017.Assessment of Atmospheric Emissions and Deposition of Major Nr Species in Indian region.2017.The Indian Nitrogen Assessment (Eds.: Y P Abrol and T K Adhya), Elsevier, pp 422–444.

Kulshrestha UC (2022a) Gaps in Nitrogen Deposition Measurements in South Asia. Current World Environment 17(2):284–288

Kulshrestha UC (2022b) Indoor Air Pollution and Reactive Nitrogen: A Serious Health Issue. Current World Environment 17(1):1–3

Kulshrestha UC (2017b) Trans-Boundary Air Pollution Suffocated the Capital. Current World Environment 12(3):465–468

Kulshrestha, U. C.(2019). Threats to Himalayan Ecosystem due to Long Range Transport of Air Pollutants and Land Use Changes. Current World Environment, 14(1), 1–2.

Kulshrestha UC, Granat L, Engardt M, Rodhe H (2005) Review of precipitation monitoring studies in India - a search for regional patterns. Atmos Environ 39(24):4419–4435

Kulshrestha UC, Jain M, Sekar R, Vairamani M, Sarkar AK, Parashar DC (2001) Chemical characteristics and source apportionment of aerosols over Indian Ocean during INDOEX-1999. Curr Sci 80:180–185

Kulshrestha UC, Kulshrestha MJ, Satyanarayana J, & Reddy, L. A. K. (2014). Atmospheric Deposition of Reactive Nitrogen in India. In Nitrogen Deposition, Critical Loads and Biodiversity (Eds: M Sutton et al), pp 75–82 (Springer).

Kulshrestha UC, Kulshrestha MJ, Sekar R, Sastry GSR, Vairamani M (2003) Chemical characteristics of rain water at an urban site of south-central India. Atmos Environ 37(21):3019–3026

Kulshrestha UC, Raman RS, Kulshrestha MJ, Rao TN, Hazarika PJ (2010) Secondary aerosol formation and identification of regional source locations by PSCF analysis in the Indo-Gangetic region of India. J Atmos Chem 63(1):33–47

Kulshrestha UC, Reddy LAK, Satyanarayana J, Kulshrestha MJ (2009) Real-time wet scavenging of major chemical constituents of aerosols and role of rain intensity in Indian region. Atmos Environ 43(32):5123–5127

Kulshrestha U (2019) Reactive nitrogen: alarming note for new fossil fuel and fertilizer policies. Curr World Environ. https://doi.org/10.12944/CWE.14.2.01

Kulshrestha UC (2013) Acid rain. In: Jorgensen SE (ed) Encyclopedia of environmental management, vol I. Taylor & Francis, New York, pp 8–22

Kulshrestha UC (2021) N-cycle and organic food: concept of modern township community farming for holistic society. Curr World Environ 16(2)

Kulshrestha UC (2023) Need to focus on nitrogen pollution. Curr World Environ 18(3). https://www.cwejournal.org/pdf/Vol18No3/CWE_Vol18_No3_p_912-913.pdf

Kulshrestha UC, Di Marco C, Nemitz E, Nissanka S, Ahmed Z, Ramachandran R, Sharma S, Rahman MM, Tshering D, Sutton MA (2024) Ammonium availability index of precipitation at South Asian sites under UKRI-GCRF South Asian nitrogen hub (AS36-A022). Paper presented at the 21st Annual Meeting of the Asia Oceania Geosciences Society, Pyeongchang-gun, Gangwon-do, South Korea

Kumar B, Singh S, Gupta GP, Lone FA, Kulshrestha UC (2016) Long range transport and wet deposition fluxes of major chemical species in snow at Gulmarg in North Western Himalayas (India). Aerosol Air Qual Res. https://doi.org/10.4209/aaqr.2015.01.0056

Kumar S, Bhatia A, Drewer J, Rees RM, Sharma S, Kumar V, Tomer R, Rahman MM, Sutton MA (2025) Mitigating ammonium and nitrate leaching in rice-wheat crop rotation: efficacy of neem coated urea and compost co-application. Front Environ Sci 13:1656231

Ladha JK, Dawe D, Pathak H, Padre AT, Yadav RL, Singh B, Singh Y, Singh Y, Singh P, Kundu AL, Sakal R (2003) How extensive are yield declines in long-term rice–wheat experiments in Asia? Field Crops Res 81(2–3):159–180

Lam SK, Suter H, Mosier AR, Chen D (2017) Using nitrification inhibitors to mitigate agricultural N_2O emission: a double-edged sword? Glob Change Biol 23(2):485–489

Linquist BA, Liu L, van Kessel C, van Groenigen KJ (2013) Enhanced efficiency nitrogen fertilizers for rice systems: Meta-analysis of yield and nitrogen uptake. Field Crop Res 154:246–254

Li P, Lu J, Wang Y, Wang S, Hussain S, Ren T, Cong R, Li X (2018) Nitrogen losses, use efficiency, and productivity of early rice under controlled-release urea. Agr Ecosyst Environ 251:78–87

Malla G, Bhatia A, Pathak H, Prasad S, Jain N, Singh J (2005) Mitigating nitrous oxide and methane emissions from soil in rice–wheat system of the Indo-Gangetic plain with nitrification and urease inhibitors. Chemosphere 58(2):141–147

Malyan SK, Bhatia A, Kumar A, Gupta DK, Singh R, Kumar SS, Tomer R, Kumar O, Jain N (2016) Methane production, oxidation and mitigation: a mechanistic understanding and comprehensive evaluation of influencing factors. Sci Total Environ 572:874–896

Malyan SK, Bhatia A, Fagodiya RK, Kumar SS, Kumar A, Gupta DK, Tomer R, Harit R, Kumar V, Jain N, Pathak H (2021) Plummeting global warming potential by chemicals interventions in irrigated rice: a lab to field assessment. Agric Ecosyst Environ 319:107545. https://doi.org/10.1016/j.agee.2021.107545

Malyan SK, Bhatia A, Kumar SS, Fagodiya RK, Pugazhendhi A, Duc PA (2019) Mitigation of greenhouse gas intensity by supplementing with Azolla and moderating the dose of nitrogen fertilizer. Biocatal Agric Biotechnol 20:101266. https://doi.org/10.1016/j.bcab.2019.101266

Martins MR, Sant'Anna SAC, Zaman M, Santos RC, Monteiro RC, Alves BJR, Jantalia CP, Boddey RM, Urquiaga S (2017) Strategies for the use of urease and nitrification inhibitors with urea: impact on N_2O and NH_3 emissions, fertilizer-15N recovery and maize yield in a tropical soil. Agric Ecosyst Environ 247:54–62. https://doi.org/10.1016/j.agee.2017.06.021

Masson-Delmotte V, Zhai P, Pirani A, Connors SL, Péan C, Berger S, Caud N, Chen Y, Goldfarb L, Gomis MI, Huang M (2021) Climate change 2021: the physical science basis. Contrib Work Group I Sixth Assess Rep Intergov Panel Climate Change 2(1):2391

Mboyerwa PA, Kibret K, Mtakwa P, Aschalew A (2022) Greenhouse gas emissions in irrigated paddy rice as influenced by crop management practices and nitrogen fertilization rates in eastern Tanzania. Front Sustain Food Syst 6:868479. https://doi.org/10.3389/fsufs.2022.868479

Mendonça EDS, Lima PCD, Guimarães GP, Moura WDM, Andrade FV (2017) Biological nitrogen fixation by legumes and N uptake by coffee plants. Rev Bras Ciênc Solo 41(00):e0160178

Miah MAM, Gaihre YK, Hunter G, Singh U, Hossain SA (2015) Fertilizer deep placement increases rice production and economic returns in southern Bangladesh. Agron J. https://doi.org/10.2134/agronj2015.0170

Mishra M, Kulshrestha UC (2021) A brief review on changes in air pollution scenario over South Asia during covid-19 lockdown. Aerosol Air Qual Res 21(4). https://doi.org/10.4209/aaqr.200541

Mishra M, Kulshrestha UC (2022) Wet deposition of total dissolved nitrogen in Indo-Gangetic Plain (India). Environ Sci Pollut Res 29(6):9282–9292. https://doi.org/10.1007/s11356-021-16293-0

Misselbrook TH, Cardenas LM, Camp V, Thorman RE, Williams JR, Rollett AJ, Chambers BJ (2014) An assessment of nitrification inhibitors to reduce nitrous oxide emissions from UK agriculture. Environ Res Lett 9(11):115006

Margon A, Parente G, Piantanida M, Cantone P, Leita L (2015) Novel investigation on ammonium thiosulphate (ATS) as an inhibitor of soil urease and nitrification. Agric Sci 6(12):1502–1512

Naseem M, Kulshrestha UC (2019) An overview of atmospheric reactive nitrogen research: South Asian perspective. Current World Environ. https://doi.org/10.12944/CWE.14.1.04

Naseem M, Kulshrestha UC (2021) Wet deposition of atmospheric inorganic reactive nitrogen (Nr) across an urban-industrial-rural transect of Nr emission hotspot (India). J Atmos Chem 78:271–304. https://doi.org/10.1007/s10874-021-09425-w

Okito A, Alves BJR, Urquiaga S, Boddey RM (2004) Nitrogen fixation by groundnut and velvet bean and residual benefit to a subsequent maize crop. Pesqui Agropecu Bras 39(12):1183–1190

Parashar DC, Kulshrestha UC, Sarma C (1998) Anthropogenic emissions of NO_x, NH_3 and N_2O in India. Nutr Cycl Agroecosyst 52:255–259

Pathak H, Bhatia A, Prasad S, Singh S, Kumar S, Jain MC, Kumar U (2002) Emission of nitrous oxide from rice-wheat systems of Indo-Gangetic Plains of India. Environ Monit Assess 77(2):163–178. https://doi.org/10.1023/A:1015823919405

Pathak H, Prasad S, Bhatia A, Singh S, Kumar S, Singh J, Jain MC (2003) Methane emission from rice–wheat cropping system in the Indo-Gangetic Plain in relation to irrigation, farmyard manure and dicyandiamide application. Agr Ecosyst Environ 97(1–3):309–316. https://doi.org/10.1016/S0167-8809(03)00033-1

Paul A, Bhatia A, Drewer J, Tomer R, Kumar V, Sharma S, Saha ND, Chakrabarti B, Shivay YS, Rees RM, Sutton MA (2025) Sulphur-coated urea reduces greenhouse gas intensity and enhances soil quality in rice cultivation. J Agric Food Res:102376

Paul A, Bhatia A, Tomer R, Kumar V, Sharma S, Pal R, Mina U, Kumar R, Manjaiah KM, Chakrabarti B, Jain N (2024) Dual inhibitors for mitigating greenhouse gas emissions and ammonia volatilization in rice for enhancing environmental sustainability. Clean Environ Syst:100199

Pugh TAM, Müller C, Elliott J, Deryng D, Folberth C, Olin S, Schmid E, Arneth A (2016) Climate analogues suggest limited potential for intensification of production on current croplands under climate change. Nat Commun 7(1):12608. https://doi.org/10.1038/ncomms12608

Rahman MM, Alam MS, Kamal MZU, Rahman GKMM, Islam MM, Haque MM, Miah MG, Biswas JC (2022a) Potential of legume-based cropping systems for climate change adaptation and mitigation. In: Meena RS, Sihag SK (eds) Advances in legumes for sustainable intensification, pp 381–402.https://doi.org/10.1016/B978-0-323-85797-0.00030-6

Rahman MM (2013) Nutrient-use and carbon-sequestration efficiencies in soils from different organic wastes in rice and tomato cultivation. Commun Soil Sci Plant Anal 44(9):1457–1471

Rahman MM (2014) Carbon and nitrogen dynamics and carbon sequestration in soils under different residue management. The Agriculturists 12:48–55

Rahman MM (2024) Soil and water: a source of life. In: Rahman MM, Biswas JC, Meena RS (eds) Climate change and soil-water-plant nexus. Springer, Singapore. https://doi.org/10.1007/978-981-97-6635-2_1

Rahman MM, Billah MM (2024) Agronomic management options for climate change adaptation and mitigation. Bangladesh J. Soil Sci 40(2):1–18

Rahman MM, Alam MS, Kamal MZU, Rahman GKMM (2020) Organic sources and tillage practices for soil management. In: Kumar S, Meena RS, Jhariya MK (eds) Resources use efficiency in agriculture. Springer, Singapore. https://doi.org/10.1007/978-981-15-6953-1_9

Rahman MM, Al-Amin HM, Alam MS, Smith J, Hillier J, Sutton MA, Adhya TK (2024) Nitrogen management options: challenges, potentials and prospects. In: Rahman MM, Biswas JC, Meena RS (eds) Climate change and soil-water-plant nexus. Springer, Singapore. https://doi.org/10.1007/978-981-97-6635-2_5

Rahman MM, Biswas JC, Sutton MA, Drewer J, Adhya TK (2021) Assessment of reactive nitrogen flows in Bangladesh's agriculture sector. Sustainability 14:272. https://doi.org/10.3390/su14010272

Rahman MM, Kamal MZ, Ranamukhaarachchi S, Alam MS, Alam MK, Khan MA, Islam MM, Alam MA, Jiban SI, Mamun MA, Abdullah HM (2022b) Effects of organic amendments on soil aggregate stability, carbon sequestration, and energy use efficiency in wetland paddy cultivation. Sustainability 14(8):4475

Rani V, Bhatia A, Kaushik R (2021) Inoculation of plant growth promoting-methane utilizing bacteria in different N-fertilizer regime influences methane emission and crop growth of flooded paddy. Sci Total Environ 775:145826

Reay DS, Davidson EA, Smith KA, Smith P, Melillo JM, Dentener F, Crutzen PJ (2012) Global agriculture and nitrous oxide emissions. Nat Clim Chang 2(6):410–416

Ritchie H, Roser M, Rosado P (2022) Fertilizers. Published online at OurWorldinData.org. https://ourworldindata.org/fertilizers [Online Resource]

Roy O, Meena RS, Kumar S, Jhariya MK, Pradhan G (2021) Assessment of land use systems for CO_2 sequestration, carbon credit potential, and income security in Vindhyan region, India. Land Degrad Dev 33(4):670–682. https://doi.org/10.1002/ldr.4181

SACEP and SANH (2021) South Asian regional cooperation on sustainable nitrogen management, nitrogen pollution in South Asia: scientific evidence, current initiatives and policy landscape. https://sanh.inms.international/publications/SACEPSANHPolicyReport

SANH (2020) Nitrogen and South Asia. https://sanh.inms.international/about/nitrogen&southasia

Sanz-Cobena A, Sánchez-Martín L, García-Torres L, Vallejo A (2012) Gaseous emissions of N_2O and NO and NO_3—leaching from urea applied with urease and nitrification inhibitors to a maize (Zeamays) crop. Agr Ecosyst Environ 149:64–73

Satyanarayana J, Reddy LAK, Kulshrestha MJ, Rao RN, Kulshrestha UC (2011) Chemical composition of rain water and influence of airmass trajectories at a rural site in an ecological sensitive area of Western Ghats (India). J Atmos Chem 66(3):101–116. https://doi.org/10.1007/s10874-011-9193-2

Soares JR, Cantarella H, de Campos Menegale ML (2012) Ammonia volatilization losses from surface-applied urea with urease and nitrification inhibitors. Soil Biol Biochem 52:82–89

Soares JR, Cantarella H, Vargas VP, Carmo JB, Martins AA, Sousa RM, Andrade CA (2015) Enhanced-efficiency fertilizers in nitrous oxide emissions from urea applied to sugarcane. J Environ Qual 44(2):423–430

Saud S, Wang D, Fahad S (2022) Improved nitrogen use efficiency and greenhouse gas emissions in agricultural soils as producers of biological nitrification inhibitors. Front Plant Sci 13:854195

Sharma A, Kulshrestha UC (2020) Wet deposition and long-range transport of major ions related to snow at Northwestern Himalayas (India). Aerosol Air Qual Res 20(5):1249–1265

Shashi MA, Mannan MA, Islam MM, Rahman MM (2018) Impact of rice husk biochar on growth, water relations and yield of maize (Zea mays L.) under drought condition. The Agriculturists 16(2):93–101

Shaviv A (2001) Advances in controlled-release fertilizers. Elsevier, pp 1–49

Shifa S, Yang A, Anik AR (2022) Bangladesh nitrogen policy report: scientific evidence, current initiatives and policy landscape. https://www.research.ed.ac.uk/en/publications/297ddb07-3f30-49db-beca-0a00a4448767

Shivay YS, Prasad R, Pal M (2016) Effect of nitrogen levels and coated urea on growth, yields and nitrogen use efficiency in aromatic rice. J Plant Nutr 39(6):875–882. https://doi.org/10.1080/01904167.2015.1109102

Shen T, Stieglmeier M, Dai J, Urich T, Schleper C (2013) Responses of the terrestrial ammonia-oxidizing archaeon Ca. *Nitrososphaeraviennensis* and the ammonia-oxidizing bacterium *Nitrosospiramultiformis* to nitrification inhibitors. FEMS Microbiol Lett 344(2):121–129

Singh B, Craswell E (2021) Fertilizers and nitrate pollution of surface and ground water: an increasingly pervasive global problem. SN Appl Sci 3(4):518. https://doi.org/10.1007/s42452-021-04521-8

Singh S, Kulshrestha UC (2012) Abundance and distribution of gaseous ammonia and particulate ammonium at Delhi (India). Biogeosciences 9:1–7.

Singh S, Bhatia A, Tomer R, Kumar V, Singh B, Singh SD (2013) Synergistic action of tropospheric ozone and carbon dioxide on yield and nutritional quality of Indian mustard (*Brassica juncea* (L.) Czern.). Environ Monit Assess 185(8):6517–6529

Singh S, Kulshrestha UC (2014) Rural versus urban gaseous inorganic reactive nitrogen in the Indo-Gangetic Plains (IGP) of India. Environ Res Lett 9(12):125004. https://doi.org/10.1088/1748-9326/9/12/125004

Singh S, Gupta GP, Kumar B, Kulshrestha UC (2014) Comparative study of indoor air pollution using traditional and improved cooking stoves in rural households of Northern India. Energy Sustain Dev 19:1–6. https://doi.org/10.1016/j.esd.2014.01.007

Singh S, Kumar B, Sharma A, Kulshrestha UC (2017) Wet deposition fluxes of atmospheric inorganic reactive nitrogen at an urban and rural site in the Indo-Gangetic Plain. Atmos Pollut Res 8(4):669–677. https://doi.org/10.1016/j.apr.2016.12.021

Smith KA (2017) Changing views of nitrous oxide emissions from agricultural soil: key controlling processes and assessment at different spatial scales. Eur J Soil Sci 68(2):137–155

Sutton MA, Drewer J, Moring A, Adhya TK, Ahmed A, Bhatia A, Brownlie W, Dragosits U, Ghude SD, Hillier J, Hooda S (2017) The Indian nitrogen challenge in a global perspective. In: The Indian nitrogen assessment. Elsevier, pp 9–28

Sutton MA, Howard CM, Erisman JW, et al (2011) The European nitrogen assessment: sources, effects and policy perspectives. Cambridge University Press. https://books.google.co.in/books?hl=en&lr=&id=oX3oFqAM9GkC&oi=fnd&pg=PR1&dq=The+European+nitrogen+assessment:+sources,+effects+and+policy+perspectives.+Cambridge+University+Press&ots=xlzlYB4o72&sig=Tsr9bDC3wA-GUhKmco_EV2hJwgs#v=onepage&q=TheEuropeannitr. Accessed 23 Feb 2021

Sutton MA, Bleeker A, Howard CM, Erisman JW, Abrol YP, Bekunda M, Datta A, Davidson E, De Vries W, Oenema O, Zhang FS (2013) Our nutrient world. The challenge to produce more food and energy with less pollution. Centre for Ecology & Hydrology, UK

Swify S, Avizienyte D, Mazeika R, Braziene Z (2022) Influence of modified urea compounds to improve nitrogen use efficiency under corn growth system. Sustainability 14(21):14166. https://doi.org/10.3390/su142114166

Saunois M, Martinez A, Poulter B, Zhang Z, Raymond PA, Regnier P, Canadell JG, Jackson RB, Patra PK, Bousquet P, Ciais P (2025) Global methane budget 2000–2020. Earth Syst Sci Data 17(5):1873–1958

Tandon HLS, Tiwari KN (2007) Fertiliser use in Indian agriculture-an eventful half century. Better Crops 1:3–5

The Business Standard (2022) Budget 2022–23: government to increase fertiliser subsidy to ensure food security. https://www.tbsnews.net/bangladesh/budget-2022-23-govt-increase-fertiliser-subsidy-ensure-food-security-424238

Thapa R, Chatterjee A, Awale R, McGranahan DA, Daigh A (2016) Effect of enhanced efficiency fertilizers on nitrous oxide emissions and crop yields: a meta-analysis. Soil Sci Soc Am J 80(5):1121–1134. https://doi.org/10.2136/sssaj2016.06.0179

Thapa R, Chatterjee A (2017) Wheat production, nitrogen transformation, and nitrogen losses as affected by nitrification and double inhibitors. Agron J 109(5):1825–1835

Timilsena YP, Adhikari R, Casey P, Muster T, Gill H, Adhikari B (2015) Enhanced efficiency fertilisers: a review of formulation and nutrient release patterns. J Sci Food Agric 95(6):1131–1142

Tiwari R, Kulshrestha U (2019) Wintertime distribution and atmospheric interactions of reactive nitrogen species along the urban transect of Delhi—NCR. Atmos Environ. https://doi.org/10.1016/j.atmosenv.2019.04.007

Tomer R, Bhatia A, Kumar V, Kumar A, Singh R, Singh B, Singh SD (2015) Impact of elevated ozone on growth, yield and nutritional quality of two wheat species in Northern India. Aerosol Air Qual Res 15(1):329–340

Trenkel T (2021) Slow-and controlled-release and Stabilized Fertilizers: an option for ME enhancing nutrient use efficiency in agriculture. International Fertilizer Industry Association (IFA)

UN (2024) Peace, dignity and equality on healthy planet. Population. https://www.un.org/en/global-issues/population. Accessed 16 Nov 2024

Vaishnav SS, Kulshrestha UC (2024) Assessmet of accumulation of Nr species through rain runoff: a case study of the Rol Village pond in Nagaur district of Rajasthan (India). Presented in N2024 conference held at New Delhi during 4–9 Feb 2024

Wrage-Mönnig N, Horn MA, Well R, Müller C, Velthof G, Oenema O (2018) The role of nitrifier denitrification in the production of nitrous oxide revisited. Soil Biol Biochem 123:A3–A16

Yadav S, Katoch A, Singh Y, Kulshrestha UC (2023) Abundance and variation of gaseous NH_3 in relation with inorganic fertilizers and soil moisture during Kharif and Rabi season. Environ Monit Assess 195(1):Article 234. https://doi.org/10.1007/s10661-022-10777-3

Yang T, Wang M, Wang X, Xu C, Fang F, Li F (2022) Product type, rice variety, and agronomic measures determined the efficacy of enhanced-efficiency nitrogen fertilizer on the CH_4 emission and rice yields in paddy fields: a meta-analysis. Agronomy 12(10):2240. https://doi.org/10.3390/agronomy12102240

Yang Y, Zhang M, Li YC, Fan X, Geng Y (2012) Controlled release urea improved nitrogen use efficiency, activities of leaf enzymes, and rice yield. Soil Sci Soc Am J 76:2307–2317. https://doi.org/10.2136/sssaj2012.0173

Yue P, Zuo X, Li K, Cui X, Wang S, Misselbrook T, Liu X (2021) The driving effect of nitrogen-related functional microorganisms under water and nitrogen addition on N_2O emission in a temperate desert. Sci Total Environ 772:145470

Zaman M, Saggar S, Blennerhassett JD, Singh J (2009) Effect of urease and nitrification inhibitors on N transformation, gaseous emissions of ammonia and nitrous oxide, pasture yield and N uptake in grazed pasture system. Soil Biol Biochem 41(6):1270–1280. https://doi.org/10.1016/j.soilbio.2009.03.011

Zaman M, Blennerhassett JD (2010) Effects of the different rates of urease and nitrification inhibitors on gaseous emissions of ammonia and nitrous oxide, nitrate leaching and pasture production from urine patches in an intensive grazed pasture system. Agr Ecosyst Environ 136(3–4):236–246

Zeng W, Li J (2020) Spatio-temporal distribution of ammonia (NH_3) emissions in agricultural fields across North China. Environ Sci Pollut Res 27:8129–8141. https://doi.org/10.1007/s11356-019-07326-w

The manufacturer's authorised representative in the EU is Springer Nature Customer Service Centre GmbH, Europaplatz 3, 69115 Heidelberg, Germany. If you have any concerns regarding our products, please contact ProductSafety@springernature.com

Printed and bound by CPI Group (UK) Ltd, Croydon, CR0 4YY
07/07/2026
02160930-0001